**National Institute of
Standards and Technology**
Technology Administration
U.S. Department of Commerce

Special Publication
800-58

Security Considerations for Voice Over IP Systems

Recommendations of the National Institute of Standards and Technology

D. Richard Kuhn, Thomas J. Walsh, Steffen Fries

NIST SP 800-58   Voice Over IP Security

NIST Special Publication 800-58

Security Considerations for Voice Over IP Systems

Recommendations of the National Institute of Standards and Technology

---

COMPUTER SECURITY

Computer Security Division
Information Technology Laboratory
National Institute of Standards and Technology
Gaithersburg, MD 20899-8930

January 2005

U.S. Department of Commerce
Donald L. Evans, Secretary

Technology Administration
Phillip J. Bond, Under Secretary for Technology

National Institute of Standards and Technology
Shashi Phoha, Director

Note to Readers

The Information Technology Laboratory (ITL) at the National Institute of Standards and Technology (NIST) promotes the U.S. economy and public welfare by providing technical leadership for the Nation's measurement and standards infrastructure. ITL develops tests, test methods, reference data, proof of concept implementations, and technical analysis to advance the development and productive use of information technology. ITL's responsibilities include the development of technical, physical, administrative, and management standards and guidelines for the cost-effective security and privacy of sensitive unclassified information in Federal computer systems.

This document is a publication of the National Institute of Standards and Technology (NIST) and is not subject to U.S. copyright. Certain commercial entities, equipment, or materials may be identified in this document in order to describe an experimental procedure or concept adequately. Such identification is not intended to imply recommendation or endorsement by the National Institute of Standards and Technology, nor is it intended to imply that the entities, materials, or equipment are necessarily the best available for the purpose.

T.J. Walsh and D.R. Kuhn are employees of NIST; S. Fries is an employee of Siemens AG.

For questions or comments on this document, contact sp800-58@nist.gov.

Acknowledgments

This document has benefited from review and comment by many experts. We particularly want to thank Mike Stauffer, of Booz Allen Hamilton, for his careful review and many contributions to improving the quality of this publication. Appendix A is derived from an internal NIST report by Tony Meehan and Tyler Moore of the University of Tulsa. Many people provided us with helpul comments and suggestions for improvements to the first draft of this document. We are grateful to Tim Grance, John Larson and colleagues at Sprint, Stephen Whitlock, David Waring, Steven Ungar, Larry L. Brock, Ron Rice, John Dabnor, Susan Landau, Cynthia Des Lauriers, Victor Marshall, Nora Wheeler, Anthony Smith, Curt Barker, Gerald Maguire, Frank Derks, Ben Halpert, Elaine Starkey, William Ryberg, Loraine Beyer, Terry Sherald, Gill Williams, Roberta Durant, Adrian Gardner, Rich Graveman, David Harrity, Lakshminath Dondeti, Mary Barnes, Cedric Aoun, Mike Lee, Paul Simmons, Marcus Leech, Paul Knight, Ken van Wyk, Manuel Vexler, and John Kelsey.

Table of Contents

Executive Summary and Recommendations 3

1 Introduction 10
    1.1   Authority 10
    1.2   Document Scope and Purpose 10
    1.3   Audience and Assumptions 11
    1.4   Document Organization 11

2 Overview of VOIP 13
    2.1   VOIP Equipment 13
    2.2   Overview of VOIP Data Handling 14
    2.3   Cost 16
    2.4   Speed and Quality 16
    2.5   Privacy and Legal Issues with VOIP 17
    2.6   VOIP Security Issues 17

3 Quality of Service Issues 19
    3.1   Latency 19
    3.2   Jitter 20
    3.3   Packet Loss 21
    3.4   Bandwidth & Effective Bandwidth 22
    3.5   The Need for Speed 24
    3.6   Power Failure and Backup Systems 24
    3.7   Quality of Service Implications for Security 25

4 H.323 26
    4.1   H.323 Architecture 26
    4.2   H.235 Security Profiles 28
        4.2.1   H.235v2 29
        4.2.2   H.235v3 32
        4.2.3   H.323 Annex J 36
        4.2.4   H.323 Security Issues 36
    4.3   Encryption Issues and Performance 37

5 SIP 39
    5.1   SIP Architecture 39
    5.2   Existing Security Features within the SIP Protocol 40
        5.2.1   Authentication of Signaling Data using HTTP Digest Authentication 41
        5.2.2   S/MIME Usage within SIP 41
        5.2.3   Confidentiality of Media Data 41
        5.2.4   TLS usage within SIP 42
        5.2.5   IPsec usage within SIP 42
        5.2.6   Security Enhancements for SIP 42
        5.2.7   SIP Security Issues 44

6 Gateway Decomposition 47
    6.1   MGCP 47
        6.1.1   Overview 47

		6.1.2	System Architecture	48
		6.1.3	Security Considerations	48
	6.2	Megaco/H.248		49
		6.2.1	Overview	49
		6.2.2	System Architecture	49
		6.2.3	Security Considerations	50

7 Firewalls, Address Translation, and Call Establishment 52

	7.1	Firewalls	52
		7.1.1 Stateful Firewalls	53
		7.1.2 VOIP specific Firewall Needs	53
	7.2	Network Address Translation	54
	7.3	Firewalls, NATs, and VOIP Issues	56
		7.3.1 Incoming Calls	56
		7.3.2 Effects on QoS	57
		7.3.3 Firewalls and NATs	57
	7.4	Call Setup Considerations with NATs and Firewalls	58
		7.4.1 Application Level Gateways	59
		7.4.2 Middlebox Solutions	59
		7.4.3 Session Border Controllers	60
	7.5	Mechanisms to solve the NAT problem	61
	7.6	Virtual Private Networks and Firewalls	62

8 Encryption & IPsec 63

	8.1	IPsec	63
	8.2	The Role of IPsec in VOIP	65
	8.3	Local VPN Tunnels	65
	8.4	Difficulties Arising from VOIPsec	65
	8.5	Encryption / Decryption Latency	66
	8.6	Scheduling and the Lack of QoS in the Crypto-Engine	67
	8.7	Expanded Packet Size	68
	8.8	IPsec and NAT Incompatibility	68

9 Solutions to the VOIPsec Issues 69

	9.1	Encryption at the End Points	69
	9.2	Secure Real Time Protocol (SRTP)	69
	9.3	Key Management for SRTP – MIKEY	71
	9.4	Better Scheduling Schemes	72
	9.5	Compression of Packet Size	72
	9.6	Resolving NAT/IPsec Incompatibilities	73

10 Planning for VOIP Deployment 75

References 78

A Appendix: VOIP Risks, Threats, and Vulnerabilities 81

	A.1	Confidentiality and Privacy	81
	A.2	Integrity Issues	83
	A.3	Availability and Denial of Service	85

B Appendix: VOIP Frequently Asked Questions 88

C Appendix: VOIP Terms 91

Index **93**

List of Figures and Tables

Figure 1. Voice Data Processing in a VOIP System. 15
Figure 2. Sample Latency Budget 20
Figure 3. H.323 Architecture 26
Figure 4. H.323 Call Setup Process 27
Table 1: H235v2 Annex D - Baseline Security Profile 29
Table 2: H235v2 Annex E – Signature Security Profile 30
Table 3: H235v2 - Voice Encryption Option 31
Table 4: H235v2 Annex F – Hybrid Security Profile 32
Table 5: H235v3 Annex D - Baseline Security Profile 33
Figure 5. SIP Network Architecture 40
Figure 6. SIP Protocol 46
Figure 7: General Scenario for MGCP Usage 48
Figure 8: General Scenario for MEGACO/H.248 Usage 50
Figure 9. IP Telephones Behind NAT and Firewall 55
Figure 10. Middlebox Communications Scenario 60
Figure 11. IPsec Tunnel and Transport Modes 64

EXECUTIVE SUMMARY AND RECOMMENDATIONS

Voice over IP – the transmission of voice over packet-switched IP networks – is one of the most important emerging trends in telecommunications. As with many new technologies, VOIP introduces both security risks and opportunities. VOIP has a very different architecture than traditional circuit-based telephony, and these differences result in significant security issues. Lower cost and greater flexibility are among the promises of VOIP for the enterprise, but VOIP should not be installed without careful consideration of the security problems introduced. Administrators may mistakenly assume that since digitized voice travels in packets, they can simply plug VOIP components into their already-secured networks and remain secure. However, the process is not that simple. This publication explains the challenges of VOIP security for agency and commercial users of VOIP, and outlines steps needed to help secure an organization's VOIP network. VOIP security considerations for the public switched telephone network (PSTN) are largely outside the scope of this document.

VOIP systems take a wide variety of forms, including traditional telephone handsets, conferencing units, and mobile units. In addition to end-user equipment, VOIP systems include a variety of other components, including call processors/call managers, gateways, routers, firewalls, and protocols. Most of these components have counterparts used in data networks, but the performance demands of VOIP mean that ordinary network software and hardware must be supplemented with special VOIP components. Not only does VOIP require higher performance than most data systems, critical services, such as Emergency 911 must be accommodated. One of the main sources of confusion for those new to VOIP is the (natural) assumption that because digitized voice travels in packets just like other data, existing network architectures and tools can be used without change. However, VOIP adds a number of complications to existing network technology, and these problems are magnified by security considerations.

Quality of Service (QoS) is fundamental to the operation of a VOIP network that meets users' quality expectations. However, the implementation of various security measures can cause a marked deterioration in QoS. These complications range from firewalls delaying or blocking call setups to encryption-produced latency and delay variation (jitter). Because of the time-critical nature of VOIP, and its low tolerance for disruption and packet loss, many security measures implemented in traditional data networks are simply not applicable to VOIP in their current form; firewalls, intrusion detection systems, and other components must be specialized for VOIP. Current VOIP systems use either a proprietary protocol, or one of two standards, H.323 and the Session Initiation Protocol (SIP). Although SIP seems to be gaining in popularity, neither of these protocols has become dominant in the market yet, so it often makes sense to incorporate components that can support both. In addition to SIP and H.323 there are also two further standards, media gateway control protocol (MGCP) and Megaco/H.248, which may be used in large deployments for gateway decomposition. These standards may be used to ease message handling with media gateways, or on the other hand they can easily be used to implement terminals without any intelligence, similar to today's phones connected to a PBX using a stimulus protocol.

Packet networks depend for their successful operation on a large number of configurable parameters: IP and MAC (physical) addresses of voice terminals, addresses of routers and firewalls, and VOIP specific software such as call processing components (call managers) and other programs used to place and route calls. Many of these network parameters are established dynamically every time network components are restarted, or when a VOIP telephone is restarted or added to the network. Because there are so many places in a network with dynamically configurable parameters, intruders have a wide array of potentially vulnerable points to attack [1].

Firewalls are a staple of security in today's IP networks. Whether protecting a LAN or WAN, encapsulating a DMZ, or just protecting a single computer, a firewall is usually the first line of defense against would be attackers. Firewalls work by blocking traffic deemed to be invasive, intrusive, or just plain malicious from flowing through them. Acceptable traffic is determined by a set of rules programmed into the firewall by the network administrator. The introduction of firewalls to the VOIP network complicates several aspects of VOIP, most notably dynamic port trafficking and call setup procedures.

Network Address Translation (NAT) is a powerful tool that can be used to hide internal network addresses and enable several endpoints within a LAN to use the same (external) IP address. The benefits of NATs come at a price. For one thing, an attempt to make a call into the network becomes very complex when a NAT is introduced. The situation is somewhat similar to an office building where mail is addressed with employees' names and the building address, but internal addressing is handled by the company mailroom. There are also several issues associated with the transmission of voice data across the NAT, including an incompatibility with IPsec. Although the use of NATs may be reduced as IPv6 is adopted, they will remain a common component in networks for years to come, so VOIP systems must deal with the complexities of NATs.

Firewalls, gateways, and other such devices can also help keep intruders from compromising a network. However, firewalls are no defense against an internal hacker. Another layer of defense is necessary at the protocol level to protect the voice traffic. In VOIP, as in data networks, this can be accomplished by encrypting the packets at the IP level using IPsec, or at the application level with secure RTP, the real-time transport protocol (RFC 3550). However, several factors, including the expansion of packet size, ciphering latency, and a lack of QoS urgency in the cryptographic engine itself can cause an excessive amount of latency in the VOIP packet delivery. This leads to degraded voice quality, again highlighting the tradeoff between security and voice quality, and emphasizing a need for speed.

VOIP is still an emerging technology, so it is difficult to develop a complete picture of what a mature worldwide VOIP network will one day look like. As the emergence of SIP has shown, new technologies and new protocol designs have the ability to radically change VOIP. Although there are currently many different architectures and protocols to choose from, eventually a true standard will emerge. Unless a widely used open standard emerges, solutions will be likely to include a number of proprietary elements, which can limit an enterprise's future choices. The most widely used of the competing standards are SIP and H.323. Some observers believe that SIP will become dominant. Major vendors are investing an increasing portion of their development effort into SIP products. An extension of SIP, the SIP for Instant Messaging and Presence Leveraging Extensions (SIMPLE) standard, is being incorporated into products that support Instant Messaging. Until a truly dominant standard emerges, organizations moving to VOIP should consider gateways and other network elements that support both H.323 and SIP. Such a strategy helps to ensure a stable and robust VOIP network in the years that come, no matter which protocol prevails.

Designing, deploying, and securely operating a VOIP network is a complex effort that requires careful preparation. The integration of a VOIP system into an already congested or overburdened network could create serious problems for the organization. There is no easy "one size fits all" solution to the issues discussed in these chapters. An organization must investigate carefully how its network is laid out and which solution fits its needs best.

NIST recommendations.

Because of the integration of voice and data in a single network, establishing a secure VOIP and data network is a complex process that requires greater effort than that required for data-only networks. In particular, start with these general guidelines, recognizing that practical considerations, such as cost or legal requirements, may require adjustments for the organization:

1. Develop appropriate network architecture.

- Separate voice and data on logically different networks if feasible. Different subnets with separate RFC 1918 address blocks should be used for voice and data traffic, with separate DHCP servers for each, to ease the incorporation of intrusion detection and VOIP firewall protection

- At the voice gateway, which interfaces with the PSTN, disallow H.323, SIP, or other VOIP protocols from the data network. Use strong authentication and access control on the voice gateway system, as with any other critical network component. Strong authentication of clients towards a gateway often presents difficulties, particularly in key management. Here, access control mechanisms and policy enforcement may help.

- A mechanism to allow VOIP traffic through firewalls is required. There are a variety of protocol dependent and independent solutions, including application level gateways (ALGs) for VOIP protocols, Session Border Controllers, or other standards-based solutions when they mature.

- Stateful packet filters can track the state of connections, denying packets that are not part of a properly originated call. (This may not be practical when multimedia protocol inherent security or lower layer security is applied, e.g., H.235 Annex D for integrity provision or TLS to protect SIP signaling.)

- Use IPsec or Secure Shell (SSH) for all remote management and auditing access. If practical, avoid using remote management at all and do IP PBX access from a physically secure system.

- If performance is a problem, use encryption at the router or other gateway, not the individual endpoints, to provide for IPsec tunneling. Since some VOIP endpoints are not computationally powerful enough to perform encryption, placing this burden at a central point ensures all VOIP traffic emanating from the enterprise network has been encrypted. Newer IP phones are able to provide Advanced Encryption System (AES) encryption at reasonable cost. Note that Federal Information Processing Standard (FIPS) 140-2, *Security Requirements for Cryptographic Modules,* is applicable to all Federal agencies that use cryptographic-based security systems to protect sensitive information in computer and telecommunication systems (including voice systems) as defined in Section 5131 of the Information Technology Management Reform Act of 1996, Public Law 104-106.

2. Ensure that the organization has examined and can acceptably manage and mitigate the risks to their information, system operations, and continuity of essential operations when deploying VOIP systems.

An especially challenging security environment is created when new technologies are deployed. Risks often are not fully understood, administrators are not yet experienced with the new technology, and security controls and policies must be updated. Therefore, agencies should carefully consider such issues as their level of knowledge and training in the technology, the maturity and quality of their security practices, controls, policies, and architectures, and their understanding of the associated security risks. These issues should be considered for all systems but are especially important with VOIP deployment for essential operations, such as systems designated "high" under FIPS 199, Standards for Security Categorization of Federal Information and Information Systems [2].

VOIP can provide more flexible service at lower cost, but there are significant tradeoffs that must be considered. VOIP systems can be expected to be more vulnerable than conventional telephone systems, in part because they are tied in to the data network, resulting in additional security weaknesses and avenues of attack (see Appendix A for more detailed discussion of vulnerabilities of VOIP and their relation to data network vulnerabilities). Confidentiality and privacy may be at greater risk in VOIP systems unless strong controls are implemented and maintained. An additional concern is the relative instability of VOIP technology compared with established telephony systems. Today, VOIP systems are still maturing and dominant standards have not emerged. This instability is compounded by VOIP's reliance on packet networks as a transport medium. The public switched telephone network is ultra-reliable. Internet service is generally much less reliable, and VOIP cannot function without Internet connections, except in the case of large corporate or other users who may operate a private network. Essential telephone services, unless carefully planned, deployed, and maintained, will be at greater risk if based on VOIP.

3. Special consideration should be given to E-911 emergency services communications, because E-911 automatic location service is not available with VOIP in some cases.

Unlike traditional telephone connections, which are tied to a physical location, VOIP's packet switched technology allows a particular number to be anywhere. This is convenient for users, because calls can be automatically forwarded to their locations. But the tradeoff is that this flexibility severely complicates the provision of E-911 service, which normally provides the caller's location to the 911 dispatch office. Although most VOIP vendors have workable solutions for E-911 service, government regulators and vendors are still working out standards and procedures for 911 services in a VOIP environment. Agencies must carefully evaluate E-911 issues in planning for VOIP deployment.

4. Agencies should be aware that physical controls are especially important in a VOIP environment and deploy them accordingly.

Unless the VOIP network is encrypted, anyone with physical access to the office LAN could potentially connect network monitoring tools and tap into telephone conversations. Although conventional telephone lines can also be monitored when physical access is obtained, in most offices there are many more points to connect with a LAN without arousing suspicion. Even if encryption is used, physical access to VOIP servers and gateways may allow an attacker to do traffic analysis (i.e., determine which parties are communicating). Agencies therefore should ensure that adequate physical security is in place to restrict access to VOIP network components. Physical securities measures, including barriers, locks, access control systems, and guards, are the first line of defense. Agencies must make sure that the proper physical countermeasures are in place to mitigate some of the biggest risks such as insertion of sniffers or other network monitoring devices. Otherwise, practically speaking this means that installation of a sniffer could result in not just data but all voice communications being intercepted.

5. Evaluate costs for additional power backup systems that may be required to ensure continued operation during power outages.

A careful assessment must be conducted to ensure that sufficient backup power is available for the office VOIP switch, as well as each desktop instrument. Costs may include electrical power to maintain UPS battery charge, periodic maintenance costs for backup power generation systems, and cost of UPS battery replacement. If emergency/backup power is required for more than a few hours, electrical generators will be required. Costs for these include fuel, fuel storage facilities, and cost of fuel disposal at end of storage life.

6. VOIP-ready firewalls and other appropriate protection mechanisms should be employed. Agencies must enable, use, and routinely test the security features that are included in VOIP systems.

Because of the inherent vulnerabilities (e.g. susceptibility to packet sniffing) when operating telephony across a packet network, VOIP systems incorporate an array of security features and protocols. Organization security policy should ensure that these features are used. Additional measures, described in this document, should be added. In particular, firewalls designed for VOIP protocols are an essential component of a secure VOIP system.

7. If practical, "softphone" systems, which implement VOIP using an ordinary PC with a headset and special software, should not be used where security or privacy are a concern.

Worms, viruses, and other malicious software are extraordinarily common on PCs connected to the internet, and very difficult to defend against. Well-known vulnerabilities in web browsers make it possible for attackers to download malicious software without a user's knowledge, even if the user does nothing more than visit a compromised web site. Malicious software attached to email messages can also be installed without the user's knowledge, in some cases even if the user does not open the attachment. These vulnerabilities result in unacceptably high risks in the use of "softphones", for most applications. In addition, because PCs are necessarily on the data network, using a softphone system conflicts with the need to separate voice and data networks to the greatest extent practical.

8. If mobile units are to be integrated with the VOIP system, use products implementing WiFi Protected Access (WPA), rather than 802.11 Wired Equivalent Privacy (WEP).

The security features of 802.11 WEP provide little or no protection because WEP can be cracked with publicly available software. The more recent WiFi Protected Access (WPA), a snapshot of the ongoing 802.11i standard, offers significant improvements in security, and can aid the integration of wireless technology with

VOIP. NIST strongly recommends that the WPA (or WEP if WPA is unavailable) security features be used as part of an overall defense-in-depth strategy. Despite their weaknesses, the 802.11 security mechanisms can provide a degree of protection against unauthorized disclosure, unauthorized network access, or other active probing attacks. However, the Federal Information Processing Standard (FIPS) 140-2, Security Requirements for Cryptographic Modules, is mandatory and binding for Federal agencies that have determined that certain information must be protected via cryptographic means. As currently defined, neither WEP nor WPA meets the FIPS 140-2 standard. In these cases, it will be necessary to employ higher level cryptographic protocols and applications such as secure shell (SSH), Transport Level Security (TLS) or Internet Protocol Security (IPsec) with FIPS 140-2 validated cryptographic modules and associated algorithms to protect information, regardless of whether the nonvalidated data link security protocols are used.

9. Carefully review statutory requirements regarding privacy and record retention with competent legal advisors.

Although legal issues regarding VOIP are beyond the scope of this document, readers should be aware that laws and rulings governing interception or monitoring of VOIP lines, and retention of call records, may be different from those for conventional telephone systems. Agencies should review these issues with their legal advisors. See Section 2.5 for more on these issues.

1 Introduction

Voice over Internet Protocol (VOIP) refers to the transmission of speech across data-style networks. This form of transmission is conceptually superior to conventional circuit switched communication in many ways. However, a plethora of security issues are associated with still-evolving VOIP technology. This publication introduces VOIP, its security challenges, and potential countermeasures for VOIP vulnerabilities.

1.1 Authority

The National Institute of Standards and Technology (NIST) developed this document in furtherance of its statutory responsibilities under the Federal Information Security Management Act (FISMA) of 2002, Public Law 107-347. NIST is responsible for developing standards and guidelines, including minimum requirements, for providing adequate information security for all agency operations and assets, but such standards and guidelines shall not apply to national security systems. This guideline is consistent with the requirements of the Office of Management and Budget (OMB) Circular A-130, Section 8b(3), "Securing Agency Information Systems," as analyzed in A-130, Appendix IV: Analysis of Key Sections. Supplemental information is provided in A-130, Appendix III.

This guideline has been prepared for use by Federal agencies. It may be used by nongovernmental organizations on a voluntary basis and is not subject to copyright, though attribution is desired.

Nothing in this document should be taken to contradict standards and guidelines made mandatory and binding on Federal agencies by the Secretary of Commerce under statutory authority, nor should these guidelines be interpreted as altering or superseding the existing authorities of the Secretary of Commerce, Director of the OMB, or any other Federal official.

1.2 Document Scope and Purpose

The purpose of this document is to provide agencies with guidance for establishing secure VOIP networks. Agencies are encouraged to tailor the recommended guidelines and solutions to meet their specific security or business requirements. VOIP security considerations for the public switched telephone network are largely outside the scope of this document. Although legal issues regarding VOIP are beyond the scope of this document, readers should be aware that laws and rulings governing interception or monitoring of VOIP lines, and retention of call records, may be different from those for conventional telephone systems. Agencies should review these issues with their legal advisors. See Section 2.5 for more on this issue.

1.3 Audience and Assumptions

VOIP is a very large, complex, and rapidly evolving field. This document reviews VOIP technologies and solutions. Each section provides background information for the reader who is new to VOIP, but most sections also include details of standards and technologies that may be of interest to technical personnel only. The following list highlights how people with different backgrounds might use this document. The intended audience includes the following:

- Managers planning to employ VOIP telephony devices in their organizations (chief information officers, senior managers, etc).

- Systems engineers and architects when designing and implementing networks.

- System administrators when administering, patching, securing or upgrading networks that include VOIP components.

- Security consultants when performing security assessments to determine security postures of VOIP environments.

- Researchers and analysts interested in VOIP technologies.

This document assumes that the readers have some minimal operating system, networking, and security expertise. Because of the rapidly changing nature of the telecommunications industry and the threats and vulnerabilities to these technologies, readers are strongly encouraged to take advantage of other resources (including those listed in this document) for more current and detailed information.

1.4 Document Organization

The document is divided into five sections followed by three appendices. This subsection is a roadmap describing the document structure.

- Section 1 is composed of an authority, purpose, scope, audience, assumptions, and document structure.

- Section 2 provides an overview of VOIP technology.

- Section 3 discusses performance and quality of service aspects of VOIP, and their effect on security options.

- Sections 4 and 5 describe the two most commonly used protocols for VOIP: H.323 and Session Initiation Protocol (SIP).

- Section 6 discusses media decomposition using the two most commonly used standards MGCP and Megaco/H.248.

- Section 7 explains requirements for call establishment and address translation in a packet-switched telephony system, and the impact of these issues on VOIP security.

- Section 8 discusses encryption technologies that can be employed in a VOIP network.

- Section 9 summarizes the options available for securing VOIP systems.

- Section 10 highlights issues to be considered in planning for VOIP.

- Appendix A describes common risks, threats, and vulnerabilities of VOIP systems, to aid system administrators in securing their systems.

- Appendix B covers some of the frequently asked questions regarding VOIP technology.

- Appendix C provides a glossary of terms and acronyms used in this document.

2 Overview of VOIP

Many readers who have a good understanding of the Internet and data communications technology may have little background in transmitting voice or real-time imaging in a packet-switched environment. One of the main sources of confusion for those new to VOIP is the (natural) assumption that because digitized voice travels in packets just like other data, existing network architectures and tools can be used without change for voice transmission. VOIP adds a number of complications to existing network technology, and these problems are compounded by security considerations. Most of this report is focused on how to overcome the complications introduced by security requirements for VOIP.

For several years, VOIP was a technology prospect, something on the horizon for the "future works" segment of telephony and networking papers. Now, however, telecommunications companies and other organizations have already, or are in the process of, moving their telephony infrastructure to their data networks. The VOIP solution provides a cheaper and clearer alternative to traditional PSTN phone lines. Although its implementation is widespread, the technology is still developing. It is growing rapidly throughout North America and Europe, but it is sometimes awkwardly implemented on most legacy networks, and often lacks compatibility and continuity with existing systems. Nevertheless, VOIP will capture a significant portion of the telephony market, given the fiscal savings and flexibility that it can provide.

2.1 VOIP Equipment

VOIP systems take a wide variety of forms. Just about any computer is capable of providing VOIP; Microsoft's NetMeeting, which comes with any Windows platform, provides some VOIP services, as does the Apple Macintosh iChat, and Linux platforms have a number of VOIP applications to choose from. In general, though, the term Voice Over IP is associated with equipment that provides the ability to dial telephone numbers and communicate with parties on the other end of a connection who have either another VOIP system or a traditional analog telephone. Demand for VOIP services has resulted in a broad array of products, including:

- Traditional telephone handset – Usually these products have extra features beyond a simple handset with dial pad. Many have a small LCD screen that may provide browsing, instant messaging, or a telephone directory, and which is also used in configuring the handset to gain access to enhanced features such as conference calls or call-park (automatic callback when a dialed number is no longer busy). Some of these units may have a "base station" design that provides the same convenience as a conventional cordless phone.

- Conferencing units – These provide the same type of service as conventional conference calling phone systems, but since communication is handled over the Internet, they may also allow users to coordinate data communication services, such as a whiteboard that displays on computer monitors at both ends.

- Mobile units – Wireless VOIP units are becoming increasingly popular, especially since many organizations already have an installed base of 802.11 networking equipment. Wireless VOIP products may present additional challenges if certain security issues are not carefully addressed. The WEP security features of 802.11b provide little or no protection. The more recent WiFi Protected Access (WPA), a snapshot of the ongoing 802.11i standard, offers significant improvements in security, and can aid the integration of wireless technology with VOIP.

- PC or "softphone" – With a headset, software, and inexpensive connection service, any PC or workstation can be used as a VOIP unit, often referred to as a "softphone". If practical, softphone systems should not be used where security or privacy are a concern. Worms, viruses, and other malicious software are common on PCs connected to the internet, and very difficult to defend against. Well known vulnerabilities in web browsers make it possible for attackers to download malicious software without a user's knowledge, even if the user does nothing more than visit a compromised web site. Malicious software attached to email messages can also be installed without the user's knowledge, in some cases even if the user does not open the attachment. These vulnerabilities result in unacceptably high risks in the use of "softphones", for most applications. In addition, because PCs are necessarily on the data network, using a softphone system conflicts with the need to separate voice and data networks to the greatest extent practical.

In addition to end-user equipment, VOIP systems include a large number of other components, including call processors (call managers), gateways, routers, firewalls, and protocols. Most of these components have counterparts used in data networks, but the performance demands of VOIP mean that ordinary network software and hardware must be supplemented with special VOIP components. The unique nature of VOIP services has a significant impact on security considerations for these networks, as will be detailed in later chapters.

2.2 Overview of VOIP Data Handling

Before any voice can be sent, a call must be placed. In an ordinary phone system, this process involves dialing the digits of the called number, which are then processed by the telephone company's system to ring the called number. With VOIP, the user must enter the dialed number, which can take the form of a number dialed on a telephone keypad or the selection of a Universal Resource Indicator (URI), but after that a complex series of packet exchanges must occur, based on a VOIP signaling protocol. The problem is that computer systems are addressed using their IP address, but the user enters an ordinary telephone number or URI to place the call. The telephone number or URI must be linked with an IP address to reach the called party, much as an alphabetic web address, such as "www.nist.gov" must be linked to the IP address of the NIST web server. A number of protocols are involved in determining the IP address that corresponds to the called party's telephone number. This process is covered in detail in Chapter 6.

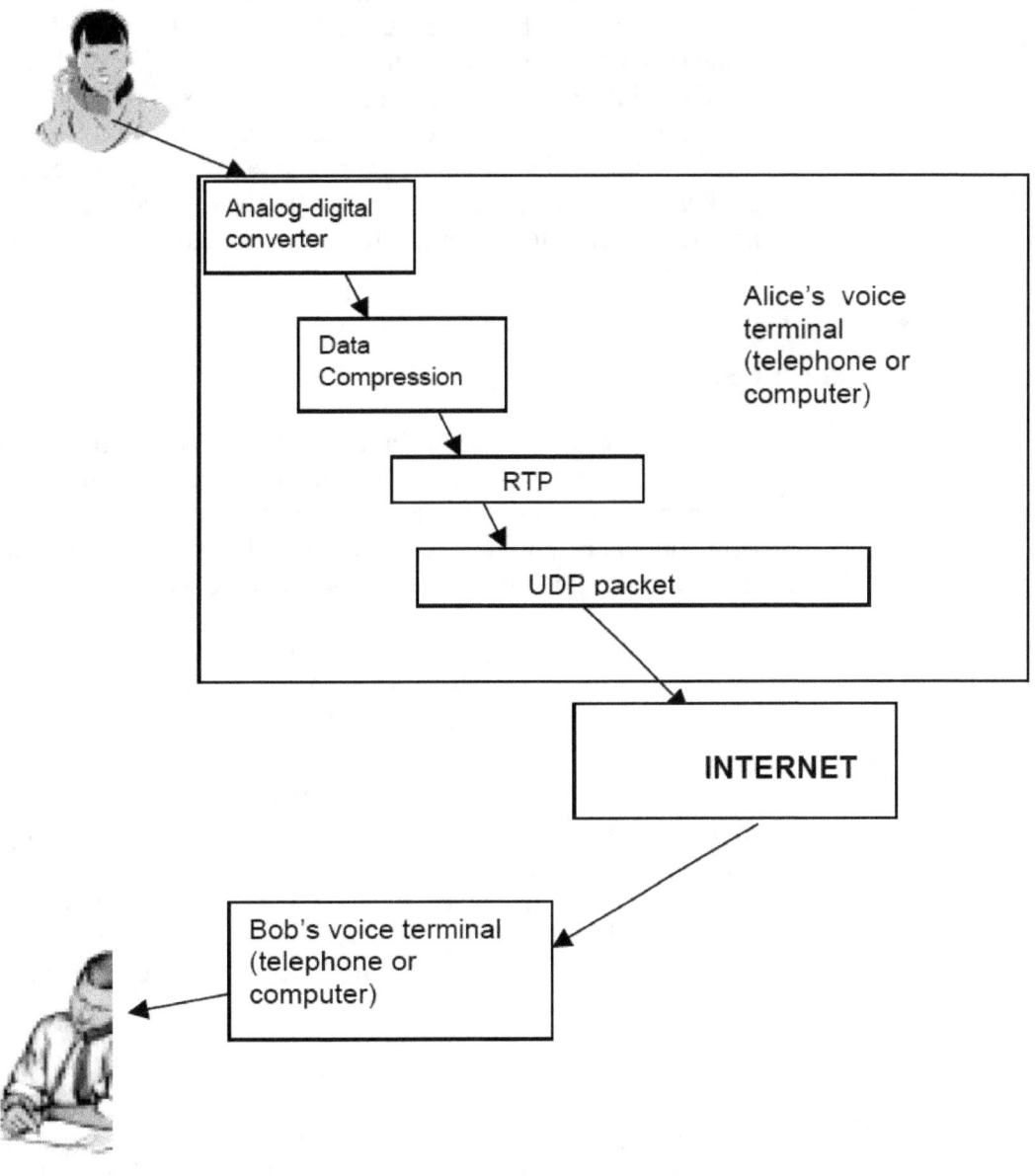

Figure 1. Voice Data Processing in a VOIP System.

Figure 1 illustrates the basic flow of voice data in a VOIP system. Once the called party answers, voice must be transmitted by converting the voice into digitized form, then segmenting the voice signal into a stream of packets. The first step in this process is converting analog voice signals to digital, using an analog-digital converter. Since digitized voice requires a large number of bits, a compression algorithm can be used to reduce the volume of data to be transmitted. Next, voice samples are inserted into data packets to be carried on the Internet. The protocol for the voice packets is typically the Real-time Transport Protocol, RTP (RFC 3550). RTP packets have special header fields that hold data needed to correctly re-assemble the packets into a voice signal on the other end. But voice packets will be carried as payload by UDP protocols that are also used for ordinary data transmission. In other words, the RTP packets are carried as data by the UDP datagrams, which can then be processed by ordinary network nodes throughout the Internet. At the other end, the process is reversed: the packets are disassembled and put into the proper order, digitized voice data extracted from the packets and uncompressed, then the digitized voice is processed by a digital-to-analog converter to render it into analog signals for the called party's handset speaker.

2.3 Cost

The feature of VOIP that has attracted the most attention is its cost-saving potential. By moving away from the public switched telephone networks, long distance phone calls become very inexpensive. Instead of being processed across conventional commercial telecommunications line configurations, voice traffic travels on the Internet or over private data network lines.

VOIP is also cost effective because all of an organization's electronic traffic (phone and data) is condensed onto one physical network, bypassing the need for separate PBX tie lines. Although there is a significant initial startup cost to such an enterprise, significant net savings can result from managing only one network and not needing to sustain a legacy telephony system in an increasingly digital/data centered world. Also, the network administrator's burden may be lessened as they can now focus on a single network. There is no longer a need for several teams to manage a data network and another to mange a voice network. The simplicity of VOIP systems is attractive, one organization / one network; but as we shall see, the integration of security measures into this architecture is very complex.

2.4 Speed and Quality

In theory, VOIP can provide reduced bandwidth use and quality superior to its predecessor, the conventional PSTN. That is, the use of high bandwidth media common to data communications, combined with the high quality of digitized voice, make VOIP a flexible alternative for speech transmission. In practice, however, the situation is more complicated. Routing all of an organization's traffic over a single network causes congestion and sending this traffic over the Internet can cause a significant delay in the delivery of speech. Also, bandwidth usage is related to digitization of voice by codecs, circuits or software processes that code and decode data for transmission. That is, producing greater bandwidth savings may slow down encoding and transmission processes. Speed and voice quality improvements are being made as VOIP networks and phones are deployed in greater numbers, and many organizations that have recently switched to a VOIP scheme have noticed no significant degradation in speed or quality.

2.5 Privacy and Legal Issues with VOIP

Although legal issues regarding VOIP are beyond the scope of this document, readers should be aware that laws and rulings governing interception or monitoring of VOIP lines may be different from those for conventional telephone systems. Privacy issues, including the security of call detail records (CDR) are addressed primarily by the Privacy Act of 1974. In addition, agencies may need to consider the Office of Management and Budget's "*Guidance on the Privacy Act Implications of Call Detail Programs to Manage Employees' Use of the Government's Telecommunication System*" (See FEDERAL REGISTER, 52 FR 12990, April 20, 1987). Because of these guidelines, many federal agencies have Privacy Act System of Record notices for the telephone CDR or usage records. CDR data may be used to reconcile the billing of services and for possible detection of waste, fraud, and abuse of government resources. In addition, NARA General Records Schedule 12, requires a 36-month retention of telephone CDR records (see http://www.archives.gov/records_management/ardor/grs12.html). VOIP systems may produce different types (and a higher volume) of CDR data than conventional telephone systems, so agencies must determine retention requirements for these records. Agencies should review any questions regarding privacy and statutory concerns with their legal advisors.

2.6 VOIP Security Issues

With the introduction of VOIP, the need for security is compounded because now we must protect two invaluable assets, our data and our voice. Federal government agencies are required by law to protect a great deal of information, even if it is unclassified. Both privacy-sensitive and financial data must be protected, as well as other government information that is categorized as sensitive but unclassified. Protecting the security of conversations is thus required. In a conventional office telephone system, security is a more valid assumption. Intercepting conversations requires physical access to telephone lines or compromise of the office private branch exchange (PBX). Only particularly security-sensitive organizations bother to encrypt voice traffic over traditional telephone lines. The same cannot be said for Internet-based connections. For example, when ordering merchandise over the phone, most people will read their credit card number to the person on the other end. The numbers are transmitted without encryption to the seller. In contrast, the risk of sending unencrypted data across the Internet is more significant. Packets sent from a user's home computer to an online retailer may pass through 15-20 systems that are not under the control of the user's ISP or the retailer. Because digits are transmitted using a standard for transmitting digits out of band as special messages, anyone with access to these systems could install software that scans packets for credit card information. For this reason, online retailers use encryption software to protect a user's information and credit card number. So it stands to reason that if we are to transmit voice over the Internet Protocol, and specifically across the Internet, similar security measures must be applied.

The current Internet architecture does not provide the same physical wire security as the phone lines. The key to securing VOIP is to use the security mechanisms like those deployed in data networks (firewalls, encryption, etc.) to emulate the security level currently enjoyed by PSTN network users. This publication investigates the attacks and defenses relevant to VOIP and explores ways to provide appropriate levels of security for VOIP networks at reasonable cost.

3 Quality of Service Issues

Quality of Service (QoS) [3,4] is fundamental to the operation of a VOIP network. Despite all the money VOIP can save users and the network elegance it provides, if it cannot deliver at least the same quality of call setup and voice relay functionality and voice quality as a traditional telephone network, then it will provide little added value. The implementation of various security measures can degrade QoS. These complications range from delaying or blocking of call setups by firewalls to encryption-produced latency and delay variation (jitter). QoS issues are central to VOIP security. If QoS was assured, then most of the same security measures currently implemented in today's data networks could be used in VOIP networks. But because of the time-critical nature of VOIP, and its low tolerance for disruption and packet loss, many security measures implemented in traditional data networks just aren't applicable to VOIP in their current form. The main QoS issues associated with VOIP that security affects are presented here:

3.1 Latency

Latency in VOIP refers to the time it takes for a voice transmission to go from its source to its destination. Ideally, we would like to keep latency as low as possible but there are practical lower bounds on the delay of VOIP. The ITU-T Recommendation G.114 [5] establishes a number of time constraints on one-way latency. The upper bound is 150 ms for one-way traffic. This corresponds to the current latency bound experienced in domestic calls across PSTN lines in the continental United States [6]. For international calls, a delay of up to 400 ms was deemed tolerable [7], but since most of the added time is spent routing and moving the data over long distances, we consider here only the domestic case and assume our solutions are upwards compatible in the international realm.

VOIP calls must achieve the 150 ms bound to successfully emulate the QoS that today's phones provide. This time constraint leaves very little margin for error in packet delivery. Furthermore, it places a genuine constraint on the amount of security that can be added to a VOIP network. The encoding of voice data can take between 1 and 30 ms [8] and voice data traveling across the North American continent can take upwards of 100 ms [9] although actual travel time is often much faster [10]. Assuming the worst case (100 ms transfer time), 20 –50 ms remain for queuing and security implementations. A less pessimistic delay budget is provided by Goode [10] and reproduced in Figure 2. (Dnw = backbone network delay, bounded to 29 ms with a total of 121 ms from other sources.)

Delay Source (G.729)	On-net Budget (ms)
Device Sample Capture	0.1
Encoding Delay (Algorithmic Delay + Processing Delay)	17.5
Packetization/ Depacketization Delay	20
Move to Output Queue/Queue Delay	0.5
Access (up) Link Transmission Delay	10
Backbone Network Transmission Delay	Dnw
Access (down) Link Transmission Delay	10
Input Queue to Application	0.5
Jitter Buffer	60
Decoder Processing Delay	2
Device Playout Delay	0.5
Total	121.1 + Dnw

Figure 2. Sample Latency Budget

Delay is not confined to the endpoints of the system. Each hop along the network introduces a new queuing delay and possibly a processing delay if it is a security checkpoint (i.e. firewall or encryption/decryption point). Also, larger packets tend to cause bandwidth congestion and increased latency. In light of these issues, VOIP tends to work best with small packets on a logically abstracted network to keep latency at a minimum.

3.2 Jitter

Jitter refers to non-uniform packet delays. It is often caused by low bandwidth situations in VOIP and can be exceptionally detrimental to the overall QoS. Variations in delays can be more detrimental to QoS than the actual delays themselves [11]. Jitter can cause packets to arrive and be processed out of sequence. RTP, the protocol used to transport voice media, is based on UDP so packets out of order are not reassembled at the protocol level. However, RTP allows applications to do the reordering using the *sequence number* and *timestamp* fields. The overhead in reassembling these packets is non-trivial, especially when dealing with the tight time constraints of VOIP.

When jitter is high, packets arrive at their destination in spurts. This situation is analogous to uniform road traffic coming to a stoplight. As soon as the stoplight turns green (bandwidth opens up), traffic races through in a clump. The general prescription to control jitter at VOIP endpoints is the use of a buffer, but such a buffer has to release its voice packets at least every 150 ms (usually a lot sooner given the transport delay) so the variations in delay must be bounded. The buffer implementation issue is compounded by the uncertainty of whether a missing packet is simply delayed an anomalously long amount of time, or is actually lost. If jitter is particularly erratic, then the system cannot use past delay times as an indicator for the status of a missing packet. This leaves the system open to implementation specific behavior regarding such a packet.

Jitter can also be controlled throughout the VOIP network by using routers, firewalls, and other network elements that support QoS. These elements process and pass along time urgent traffic like VOIP packets sooner than less urgent data packets. However, not all network components were designed with QoS in mind. An example of a network element that does not implement this QoS demand is a crypto-engine that ignores Type of Service (ToS) bits in an IP header and other indicators of packet urgency (see Section 8.6). Another method for reducing delay variation is to pattern network traffic to diminish jitter by making as efficient use of the bandwidth as possible. This constraint is at odds with some security measures in VOIP. Chief among these is IPsec, whose processing requirements may increase latency, thus limiting effective bandwidth and contributing to jitter. Effective bandwidth is compromised when packets are expanded with new headers. In normal IP traffic, this problem is negligible since the change in the size of the packet is very small compared with the packet size. Because VOIP uses very small packets, even a minimal increase is important because the increase accrues across all the packets, and VOIP sends a very high volume of these small packets.

The window of delivery for a VOIP packet is very small, so it follows that the acceptable variation in packet delay is even smaller. Thus, although we are concerned with security, the utmost care must be given to assuring that delays in packet deliveries caused by security devices are kept uniform throughout the traffic stream. Implementing devices that support QoS and improving the efficiency of bandwidth with header compression allows for more uniform packet delay in a secured VOIP network.

3.3 Packet Loss

VOIP is exceptionally intolerant of packet loss. Packet loss can result from excess latency, where a group of packets arrives late and must be discarded in favor of newer ones. It can also be the result of jitter, that is, when a packet arrives after its surrounding packets have been flushed from the buffer, making the received packet useless. VOIP-specific packet loss issues exist in addition to the packet loss issues already associated with data networks; these are the cases where a packet is not delivered at all. Compounding the packet loss problem is VOIP's reliance on RTP, which uses the unreliable UDP for transport, and thus does not guarantee packet delivery. However, the time constraints do not allow for a reliable protocol such as TCP to be used to deliver media. By the time a packet could be reported missing, retransmitted, and received, the time constraints for QoS would be well exceeded. The good news is that VOIP packets are very small, containing a payload of only 10-50 bytes [8], which is approximately 12.5-62.5 ms, with most implementations tending toward the shorter range. The loss of such a minuscule amount of speech is not discernable or at least not worthy of complaint for a human VOIP user. The bad news is these packets are usually not lost in isolation. Bandwidth congestion and other such causes of packet loss tend to affect all the packets being delivered around the same time. So although the loss of one packet is fairly inconsequential, probabilistically the loss of one packet means the loss of several packets, which severely degrades the quality of service in a VOIP network.

In a comparison of VOIP quality versus traditional circuit switched networks, Sinden [12] reported data from a Telecommunications Industry Association (TIA) study that showed even a fairly small percentage of lost packets could push VOIP network QoS below the level users have come to expect on their traditional phone lines. Each codec the TIA studied experienced a steep downturn in user satisfaction when latency crossed the 150 ms point. However, even with less than 150 ms of latency, a packet loss of 5% caused VOIP traffic encoded with G.711 (an international standard for encoding telephone audio on a 64 kbps stream) to drop below the QoS levels of the PSTN, even with a packet loss concealment scheme. Similarly, losses of 1 and 2 percent, respectively, were enough to place quality in VOIP networks encoded with G.723.1 (for very low bit rate speech compression) and G.729A (for voice compression on an 8kbps stream) below this threshold. At losses of 3 and 4 percent, respectively, the performance of these networks resulted in a majority of dissatisfied users. These results corroborated the findings of a 1998 study at UCal-Berkeley that determined "tolerable loss rates are within 1-3% and the quality becomes intolerable when more than 3% of the voice packets are lost" [9]. Both studies found that greater payload compression rates resulted in a higher sensitivity to packet loss. On the bright side, the implementation of forward error correction [9] and packet loss concealment schemes produced a VOIP network that was less sensitive to packet loss. The percentages presented in both studies did not take into account varying packet sizes and several other properties that can affect the relationship between packet loss and QoS.

Despite the infeasibility of using a guaranteed delivery protocol such as TCP, there are some remedies for the packet loss problem. One cannot guarantee all packets are delivered, but if bandwidth is available, sending redundant information can probabilistically annul the chance of loss. Such bandwidth is not always accessible and the redundant information will have to be processed, introducing even more latency to the system and ironically, possibly producing even greater packet loss. Newer codecs such as internet Low Bit-rate Codec (iLBC) are also being developed that offer roughly the voice quality and computational complexity of G.729A, while providing increased tolerance to packet loss.

3.4 Bandwidth & Effective Bandwidth

In any network, the obvious first concern is whether the network is available for use. Since a network can be broken down into nodes and links between nodes where traffic flows, the quest for an available network boils down to the availability of each node, and the availability of each path between the nodes. Later on, we will consider the nodes themselves, in cases where firewalls, CPUs, or other endpoints are unavailable, but for now we will concentrate on the availability of the edges: the bandwidth of the VOIP system.

As in data networks, bandwidth congestion can cause packet loss and a host of other QoS problems. Thus, proper bandwidth reservation and allocation is essential to VOIP quality. One of the great attractions of VOIP, data and voice sharing the same wires, is also a potential headache for implementers who must allocate the necessary bandwidth for both networks in a system normally designed for one. Congestion of the network causes packets to be queued, which in turn contributes to the latency of the VOIP system. Low bandwidth can also contribute to non-uniform delays (jitter), since packets will be delivered in spurts when a window of opportunity opens up in the traffic.

Because of these issues, VOIP network infrastructures must provide the highest amount of bandwidth possible. On a LAN, this means having modern switches running at 100M bit/sec and other architectural upgrades that will alleviate bottlenecks within the LAN. Percy and Hommer [13] suggest that if network latencies are kept below 100 milliseconds, maximum jitter never more than 40 milliseconds, then packet loss should not occur. With these properties assured, one can calculate the necessary bandwidth for a VOIP system on the LAN in a worst case scenario using statistics associated with the worst-case bandwidth congesting codec [13]. This is fine when dealing simply with calls across the LAN, but the use of a WAN complicates matters. Bandwidth usage varies significantly across a WAN, so a much more complex methodology is needed to estimate required bandwidth usage. Chuah [9] provides an analysis of the aggregate bandwidth needed in terms of the amount of traffic and its rate of flow.

Methods for reducing the bandwidth usage of VOIP include RTP header compression and Voice Activity Detection (VAD). RTP compression condenses the media stream traffic so less bandwidth is used. However, an inefficient compression scheme can cause latency or voice degradation, causing an overall downturn in QoS. VAD prevents the transmission of empty voice packets (i.e. when a user is not speaking, their device does not simply send out white noise). However, by definition VAD will contribute to jitter in the system by causing irregular packet generation.

The bandwidth requirements put forth by Chuah are designed for a basic VOIP system. Adding security constraints significantly increases the bandwidth usage, causing more latency and jitter, thereby degrading the overall QoS of the network. In addition, these requirements do not explicitly take into account the heterogeneous data flow over the network. Since voice and data streams are sharing the same finite bandwidth, and data streams tend to contain much larger packets than VOIP, significant amounts of data can congest the network and prevent voice traffic from reaching its destination in a timely fashion. For this reason, most new hardware devices deployed on networks support QoS for VOIP. These devices, such as routers and firewalls, make use of the IP protocol's Type of Service (ToS) bits to send VOIP traffic through before less time urgent data traffic. VOIP phones often also include QoS features [13].

Not only is the available bandwidth of the system affected by the introduction of security measures, but in addition the *effective bandwidth* of the VOIP system is significantly depreciated. Effective bandwidth is defined by Barbieri et al. [8] as "the percentage of bandwidth carrying actual data with regard to the total bandwidth used." The introduction of IPsec or other forms of encryption results in a much larger header to payload ratio for each packet, and this reduces the effective bandwidth as the same number of packets (but larger sized) are used to transport the same amount of data. The consequences of this reduction include decreased throughput and increased latency.

3.5 The Need for Speed

The key to conquering QoS issues like latency and bandwidth congestion is speed. By definition, faster throughput means reduced latency and probabilistically reduces the chances of severe bandwidth congestion. Thus every facet of network traversal must be completed quickly in VOIP. The latency often associated with tasks in data networks will not be tolerated. Chief among these latency producers that must improve performance are firewall/NAT traversal and traffic encryption/decryption. Traditionally, these are two of the most effective ways for administrators to secure their networks. However, they are also two of the greatest contributors to network congestion and throughput delay. Inserting traditional firewall and encryption products into a VOIP network is not feasible, particularly when VOIP is integrated into existing data networks. Instead, these data-network solutions must be adapted to support security in the new fast paced world of VOIP. The next several chapters explore the resolution of the conflict between the speed demands of QoS and the slowdown associated with these traditional security measures.

3.6 Power Failure and Backup Systems

Conventional telephones operate on 48 volts supplied by the telephone line itself. This is why home telephones continue to work even during a power failure. Most offices use PBX systems with their conventional telephones, and PBXs require backup power systems so that they continue to operate during a power failure. These backup systems will continue to be required with VOIP, and in many cases will need to be expanded. An organization that provides uninterruptible power systems for its data network and desktop computers may have much of the power infrastructure needed to continue communication functions during power outages, but a careful assessment must be conducted to ensure that sufficient backup power is available for the office VOIP switch, as well as each desktop instrument. Costs may include electrical power to maintain UPS battery charge, periodic maintenance costs for backup power generation systems, and cost of UPS battery replacement. If emergency/backup power is required for more than a few hours, electrical generators will be required. Costs for these include fuel, fuel storage facilities, and cost of fuel disposal at end of storage life.

3.7 Quality of Service Implications for Security

The strict performance requirements of VOIP have significant implications for security, particularly denial of service (DoS) issues [14]. VOIP-specific attacks (i.e., floods of specially crafted SIP messages) may result in DoS for many VOIP-aware devices. For example, SIP phone endpoints may freeze and crash when attempting to process a high rate of packet traffic SIP proxy servers also may experience failure and intermittent log discrepancies with a VOIP-specific signaling attack of under 1Mb/sec. In general, the packet rate of the attack may have more impact than the bandwidth; i.e., a high packet rate may result in a denial of service even if the bandwidth consumed is low.

4 H.323

H.323 is the ITU specification for audio and video communication across packetized networks. H.323 is actually an umbrella standard, encompassing several other protocols, including H.225, H.245, and others. It acts as a wrapper for a suite of media control recommendations by the ITU. Each of these protocols has a specific role in the call setup process, and all but one are made to dynamic ports. Figure 3 shows the H.323 architecture and Figure 4 provides an overview of the H.323 call setup process.

4.1 H.323 Architecture

An H.323 network is made up of several endpoints (terminals), a gateway, and possibly a gatekeeper, Multipoint control unit, and Back End Service. The gatekeeper is often one of the main components in H.323 systems. It provides address resolution and bandwidth control. The gateway serves as a bridge between the H.323 network and the outside world of (possibly) non-H.323 devices. This includes SIP networks and traditional PSTN networks. This brokering can add to delays in VOIP, and hence there has been a movement towards the consolidation of at least the two major VOIP protocols [15]. A Multipoint Control Unit is an optional element that facilitates multipoint conferencing and other communications between more than two endpoints. Gatekeepers are an optional but widely used component of a VOIP network [16]. If a gatekeeper is present, a Back End Service (BES) may exist to maintain data about endpoints, including their permissions, services, and configuration [17].

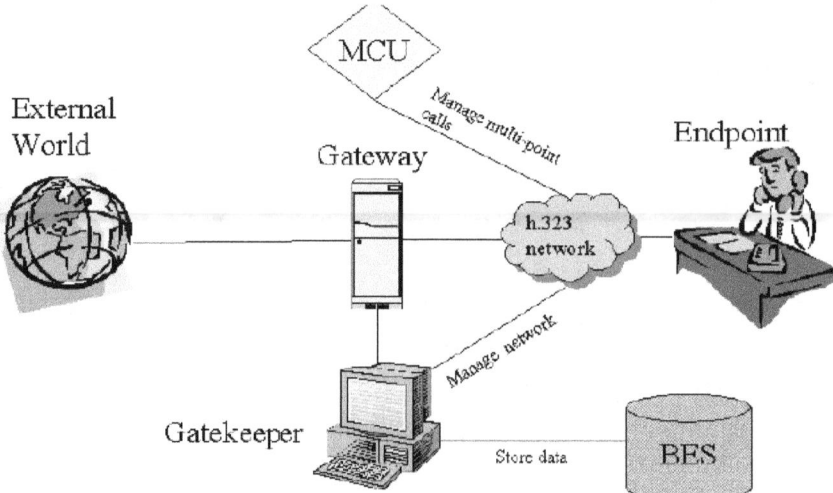

Figure 3. H.323 Architecture

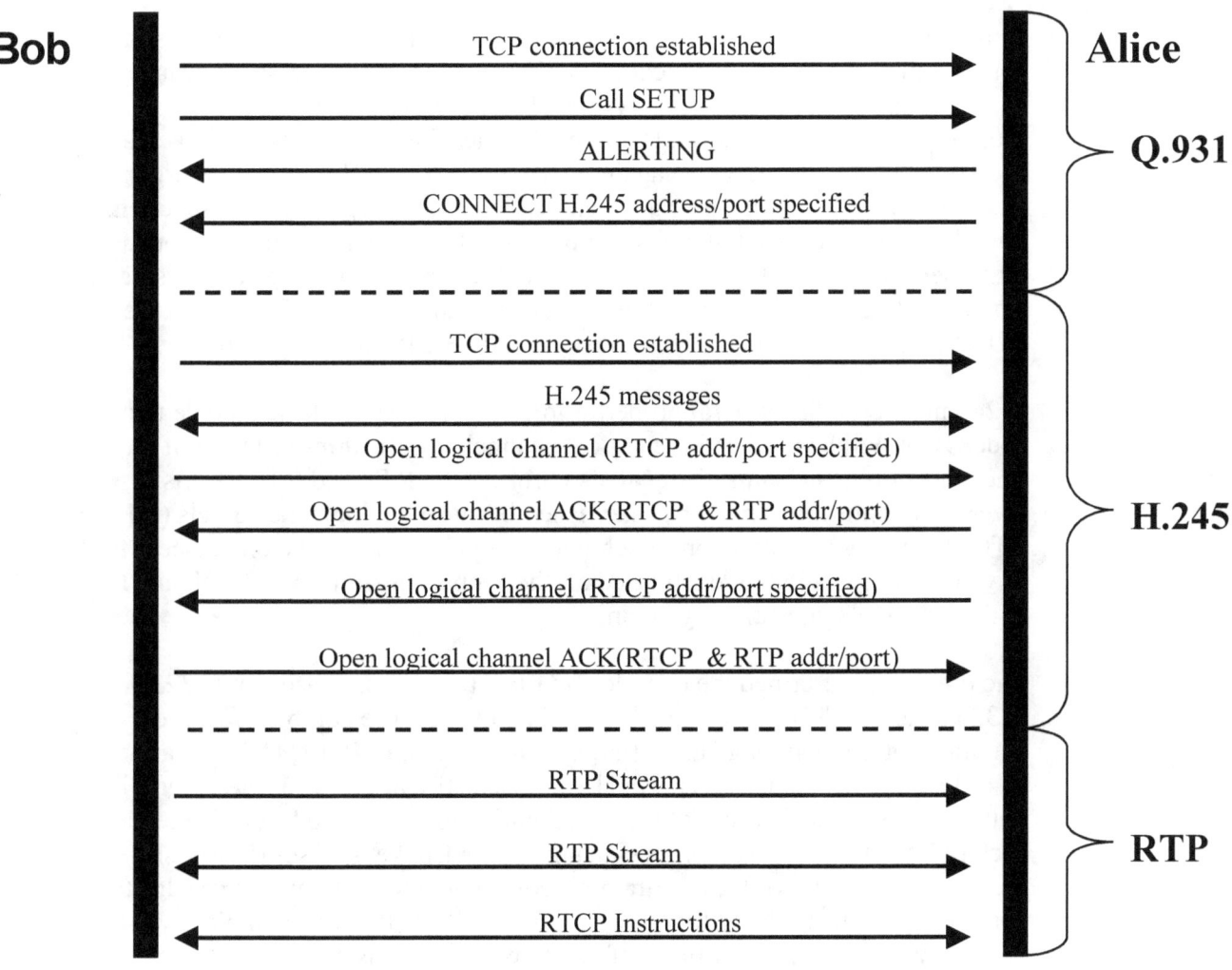

Figure 4. H.323 Call Setup Process

Generally, there are different types of H.323 calls defined in the H.323 standard:
- Gatekeeper routed call with gatekeeper routed H.245 signaling
- Gatekeeper routed call with direct H.245 signaling
- Direct routed call with gatekeeper
- Direct routed call without gatekeeper

An H.323 VOIP session is initiated (depending on the call model used) by either a TCP or a UDP (if RAS is the starting point) connection with an H.225 signal. In the case of UDP this signal contains the Registration Admission Status (RAS) protocol that negotiates with the gatekeeper and obtains the address of the endpoint it is attempting to contact. Then a "Q.931-like" protocol (still within the realm of H.225) is used to establish the call itself and negotiate the addressing information for the H.245 signal. (This is done via TCP; Q.931 actually encapsulates the H.225 Call Signaling messages.) This "setup next" procedure is common throughout the H.323 progression where one protocol negotiates the configuration of the next protocol used. In this case, it is necessary because H.245 has no standard port [10]. While H.225 simply negotiates the establishment of a connection, H.245 establishes the channels that will actually be used for media transfer. Once again, this is done over TCP. In a time-urgent situation, the H.245 message can be embedded within the H.225 message (H.245 tunneling), but the speed of a call setup is usually a QoS issue that vendors and customers are willing to concede for better call quality. H.323 also offers Fast Connect. Here, a call may be setup using one roundtrip. The SETUP and the CONNECT messages piggyback the necessary H.245 signaling elements.

H.245 must establish several properties of the VOIP call. These include the audio codecs that will be used and the logical channels for the transportation of media. The "OpenLogicalChannel" signal also brokers the RTP and RTCP ports. Overall, 4 connections must be established because the logical channels (RTP and RTCP) are only one direction. Each one-way pair must also be on adjacent ports as well. After H.245 has established all the properties of the VOIP call and the logical channels, the call may begin.

The preceding described the complex VOIP setup process based on H.323.. The H.323 suite has different protocols associated with more complex forms of communication including H.332 (large conferences), H.450.1, H.450.2, and H.450.3 (supplementary services), H.235 (security), and H.246 (interoperability with circuit switched services) [18]. Authentication may also be performed at each point in the call setup process using symmetric keys or some prior shared secret [19]. The use of these extra protocols and/or security measures adds to the complexity of the H.323 setup process. We shall see that this complexity is paramount in the incompatibility of H.323 with firewalls and NATs. These issues are discussed at length in the next section.

4.2 H.235 Security Profiles

With the establishment of the H.235 version 2 standard in November 2000 the ITU-T took a step towards interoperability by defining different security profiles. This was necessary because the standard itself does not mandate particular features. The defined profiles provide different levels of security and describe a subset of possible security mechanisms offered by H.235. They comprise different options for the protection of communications, e.g., by using different options of H.235, which results in different implementation impact. The following subsections provide a short overview of the profiles provided by different organizations.

4.2.1   H.235v2

H.235v2 is the followup version of H.235 that was approved in November 2000. Besides enhancements such as the support of elliptic curve cryptography and the support for the Advanced Encryption System (AES) standard, several security profiles are defined to support product interoperability. These profiles are defined in annexes to H.235v2 as follows:
- Annex D – Shared secrets and keyed hashes
- Annex E – Digital signatures on every message
- Annex F – Digital signatures and shared secret establishment on first handshake, afterwards keyed hash usage

4.2.1.1   H.235v2 Annex D – Baseline Security Profile

The Baseline Security Profile relies on symmetric techniques. Shared secrets are used to provide authentication and/or message integrity. The supported scenarios for this profile are endpoint to gatekeeper, gatekeeper to gatekeeper, and endpoint to endpoint. For the profile the gatekeeper-routed signaling (hop-by-hop security) is favored. Using it for the direct call model is generally possible but limited due to the fact that a shared secret has to be established between the parties that want to communicate before the actual communication takes place. This might be possible in smaller environments but may lead to a huge administrative effort in larger environments.

This profile supports secure fast connect and H.245 tunneling, and may be combined with the Voice Encryption Option described in section 4.2.1.3. Note: that this profile is easy to implement but it is not really scalable for "global" IP telephony due to the restricted key management.

Security Services	Call Functions			
	RAS	H.225.0	H.245	RTP
Authentication	Shared Secret (Password), HMAC-SHA1-96	Shared Secret (Password), HMAC-SHA1-96	Shared Secret (Password), HMAC-SHA1-96	
Access Control				
Non-Repudiation				
Confidentiality				
Integrity	Shared Secret (Password), HMAC-SHA1-96	Shared Secret (Password), HMAC-SHA1-96	Shared Secret (Password), HMAC-SHA1-96	
Key Management	Subscription-based password assignment	Subscription-based password assignment		

Table 1: H235v2 Annex D - Baseline Security Profile

4.2.1.2 H.235v2 Annex E – Signature Security Profile

The Signature Security Profile relies on asymmetric techniques. Certificates and digital signatures are used to provide authentication and message integrity. The signature security profile mandates the gatekeeper-routed model. Other call models are for further study. Since this profile relies on a public key infrastructure rather than on pre-established shared secrets it scales for larger, global environments. In addition to the Baseline Security Profile it provides non-repudiation.

Security Services	Call Functions			
	RAS	H.225.0	H.245	RTP
Authentication	SHA1/ MD5, digital signature	SHA1/MD5, digital signature	SHA1/MD5, digital signature	
Access Control				
Non-Repudiation	SHA1/ MD5, digital signature	SHA1/ MD5, digital signature	SHA1/ MD5, digital signature	
Confidentiality				
Integrity	SHA1/ MD5, digital signature	SHA1/ MD5, digital signature	SHA1/ MD5, digital signature	
Key Management	certificate allocation	certificate allocation		

Table 2: H235v2 Annex E – Signature Security Profile

This profile supports secure fast connect and H.245 tunneling and may be combined with the Voice Encryption Option described in section 4.2.1.3. Note: This protocol may have a critical impact on overall performance. This is due to the use of digital signatures for every message, requiring signature generation and verification on the sender's and the receiver's side. The Hybrid Security Profile described in section 4.2.1.4 provides an alternative to the Signature Security Profile.

4.2.1.3 H.235v2 Annex D - Voice Encryption Option

The voice encryption option offers confidentiality for the voice media stream data and may be combined with the baseline or the signature security profile.

Security Services	Call Functions			
	RAS	H.225.0	H.245	RTP
Authentication				
Access Control				
Non-Repudiation				
Confidentiality				56 bit DES or 56- RC2®/ 168-bit

	RAS	H.225.0	H.245	RTP
				Triple-DES, AES
Integrity				
Key Management		Authenticated Diffie-Hellman key agreement	Integrated H.235 session key management (key distribution, key update); certificate requests	

Table 3: H235v2 - Voice Encryption Option

The voice encryption option describes the master key exchange during H.225.0 call signaling and the generation and distribution of media stream keys during H.245 call control. The encryption algorithms are to be used in CBC mode. New is the support of the AES. AES and TDEA may also be used in EOFB mode.

The following security mechanisms are described within the voice encryption security profile:
- Encryption of RTP packets with an assortment of algorithms and modes to be taken;
- Key management with key and security capability exchange;
- Key update mechanism and synchronization.

The following issues are not covered by this profile:
- Encryption and key management for RTCP;
- Authentication and integrity for RTP and RTCP (a lightweight authentication and integrity could be provided by media anti-spamming).

To counter denial of service and flooding attacks on discovered RTP/UDP ports, the H.235 standard defines the media anti-spamming procedure, which provides lightweight RTP packet authentication and integrity on selected fields through a computed message authentication code (MAC). The algorithms used are triple-DES-MAC or the cryptographic one-way function SHA1. Media anti-spamming uses the padding mechanism of RTP. For this feature, no special security profile was specified in H.235 like the voice encryption security profile for the RTP encryption, but media anti-spamming may be used in combination with media encryption.

4.2.1.4 H.235v2 Annex F – Hybrid Security Profile

The Hybrid Security Profile relies on asymmetric and symmetric techniques. It can be seen as a combination of the Baseline and the Signature Security Profile. Certificates and digital signatures are used to provide authentication and message integrity (as in the Signature Security Profile) for the first handshake between two entities. During this handshake a shared secret is established that will be used further on in the same way described for the Baseline Security Profile. The hybrid security profile mandates the gatekeeper-routed model. Other call models are open for further study.

Since this profile relies on a public key infrastructure rather than on pre-established shared secrets it scales for larger, global environments.

This profile supports secure fast connect and H.245 tunneling and may be combined with the Voice Encryption Option described in section 4.2.1.3. Note: This profile provides high security without relying on pre-established shared secrets. Due to the key management using digital signatures it is scalable for "global" IP telephony. Moreover, it does not suffer from the same performance requirements as the Signature Security Profile described in section 4.2.1.2.

Security Services	Call Functions			
	RAS	H.225.0	H.245	RTP
Authentication	RSA digital signature, (SHA1)	RSA digital signature, (SHA1)	RSA digital signature, (SHA1)	
	HMAC-SHA1-96	HMAC-SHA1-96	HMAC-SHA1-96	
Access Control				
Non-Repudiation	Only for first handshake send between two entities	Only for first handshake send between two entities		
Confidentiality				
Integrity	RSA digital signature, (SHA1)	RSA digital signature, (SHA1)	RSA digital signature, (SHA1)	
	HMAC-SHA1-96	HMAC-SHA1-96	HMAC-SHA1-96	
Key Management	certificate allocation	certificate allocation		
	authenticated Diffie-Hellman key agreement	authenticated Diffie-Hellman key agreement		

Table 4: H235v2 Annex F – Hybrid Security Profile

4.2.2 H.235v3

Version 3 of H.235 supersedes H.235 version 2 featuring a procedure for encrypted DTMF (touch tone) signals, object identifiers for the AES encryption algorithm for media payload encryption, and the Enhanced Outer FeedBack (EOFB) stream-cipher encryption mode for encryption of media streams. Moreover, an authentication-only option in Annex D for smooth NAT/firewall traversal is introduced as well as better security support for direct-routed calls in a new Annex I. Error reporting is also improved. Annex G is also discussed to support H.235v3. Annex G describes a profile to support SRTP.

4.2.2.1 H.235 Annex D – Baseline Security Profile Enhancements

Using this profile, either message authentication and integrity is achieved by calculating an integrity check value over the complete message, or authentication only by computing an integrity check over a special part of the message. The latter option is useful in environments where NAT and Firewalls are applied. The version used is distinguished by an identifier.

Security Services	Call Functions			
	RAS	H.225.0	H.245	RTP
Authentication	Shared Secret (Password), HMAC-SHA1-96	Shared Secret (Password), HMAC-SHA1-96	Shared Secret (Password), HMAC-SHA1-96	
Access Control				
Non-Repudiation				
Confidentiality				
Integrity (optional)	*Shared Secret (Password), HMAC-SHA1-96*	*Shared Secret (Password), HMAC-SHA1-96*	*Shared Secret (Password), HMAC-SHA1-96*	
Key Management	Subscription-based password assignment	Subscription-based password assignment		

Table 5: H235v3 Annex D - Baseline Security Profile

Table 5 shows the updated version of the baseline security profile, where the integrity protection of the signaling data is marked as optional.

4.2.2.2 Draft H.235v3 Annex G – SRTP & MIKEY usage

Annex G discusses the incorporation of key management supporting the Secure Real-time Transport Protocol (SRTP). SRTP provides confidentiality, message authentication and replay protection to the RTP/RTCP traffic. The RTP standard provides the flexibility to adapt to application specific requirements with the possibility to define profiles in companion documents. SRTP is defined as such a profile of the RTP protocol and is currently stardards-track RFC 3711. SRTP may be used within multimedia sessions to ensure a secure media data exchange. It can be used with several session control protocols, e.g., with H.323 or SIP.

SRTP does not define key management by itself. It rather uses a set of negotiated parameters from which session keys for encryption, authentication and integrity protection are derived. The key management is not fixed. Within the IETF, the MSEC working group discusses key management solutions to be used beyond other protocols with SRTP. The preferred solution here is Multimedia Internet Keying (MIKEY) [RFC 3830] which is also part of the group key management architecture (GKMArch).

MIKEY describes a key management scheme that addresses real-time multimedia scenarios (e.g. SIP calls and RTSP sessions, streaming, unicast, groups, multicast). The focus lies on the setup of a security association for secure multimedia sessions including key management and update, security policy data, etc., such that requirements in a heterogeneous environment are fulfilled. MIKEY also supports the negotiation of single and multiple crypto sessions. This is especially useful for the case where the key management is applied to SRTP, since here RTP and RTCP may to be secured independently. Deployment scenarios for MIKEY comprise peer-to-peer, simple one-to-many, and small-size interactive group scenarios.

MIKEY supports the negotiation of cryptographic keys and security parameters (SP) for one or more security protocols. This results in the concept of crypto session bundles, which describe a collection of crypto sessions that may have a common Traffic Encryption Key (TEK), Generation Key (TGK), and session security parameters.

MIKEY defines three options for the user authentication and negotiation of the master keys all as 2 way-handshakes. They are:

- Symmetric key distribution (pre-shared keys, MAC for integrity protection)
- Asymmetric key distribution
- Diffie Hellman key agreement protected by digital signatures

A fourth version exists, which is not part of MIKEY itself. It is specified as an extension to MIKEY and describes the Diffie-Hellman key agreement protected by symmetric pre-shared keys.

The default and mandatory key transport encryption is AES in counter mode. MIKEY uses a 160-bit authentication tag, generated by HMAC with SHA-1 as the mandatory algorithm as described in RFC 2104. Also mandatory, when asymmetric mechanisms are used, is the support of X.509v3 certificates for public key encryption and digital signatures.

Annex G discusses the use of MIKEY to integrate key management suitable for SRTP in three profiles:
- Profile 1 using symmetric techniques to protect the key management data in gatekeeper routed scenarios;
- Profile 2 using asymmetric techniques to protect the key management data in scenarios with a single gatekeeper instance;

- Profile 3 describes Profile 2 for multiple intermediate gatekeepers. The basic concept of all profiles is the protected transmission of the key management data as self-contained container.

4.2.2.3 Draft H.235v3 Annex H – RAS Key Management

The basic idea formulated in H.235 Annex H is key management negotiation during the RAS gatekeeper discovery phase. During gatekeeper discovery a shared secret is established between the endpoint and the gatekeeper. The negotiation of the shared secret may be protected using PINs or passwords during the initial phase of the protocol.

The draft references two protocols for Encrypted Key Exchange using a shared secret to "obscure" a Diffie-Hellman key exchange. The first one is the Encrypted Key Exchange (EKE), where the shared secret is used to encrypt the Diffie-Hellman public keys under a symmetric algorithm. The second one is the Simple Password-authenticated Exponential Key Exchange (SPEKE) method [20], where the shared secret builds a generator for the Diffie-Hellman group. The usage of these protocols leads to a strong Diffie-Hellman key exchange with use of the shared secret. A potential disadvantage of these protocols is that they are typically subject to patent protection.

The draft discusses the use of the PIN or password for the protection of the exchange of the public parameter of public key system (Diffie Hellman, elliptic curves) by encryption using a symmetric algorithm in CBC mode. To be more specific, the password or PIN is used to derive the initialization vectors for the encryption algorithms. The negotiated keys and algorithms may then be applied later on to protect the further RAS and call signaling phase.

One option to protect the call signaling phase is TLS, which is discussed further in the draft Annex H. Here, the RAS negotiation replaces the initial TLS handshake protocol. This is obviously only useful if the call signaling is gatekeeper routed. The approach is especially useful for inter-gatekeeper authentication and signaling using the LRQ/LCF exchange. In this case, there is no third RAS message by which the calling gatekeeper can authenticate itself to the called gatekeeper using the negotiated key material, but the caller can be implicitly authenticated by its ability to establish the call signaling channel with the correct TLS session parameters. TLS can then be deployed without the costly handshake phase using only the recode layer of TLS together with the negotiated key material and algorithms from the RAS phase.

4.2.2.4 H.235v3 Annex I – H.235 Annex D for Direct Routed Scenarios

Both Annex D and Annex F are to be used in gatekeeper routed environments. Annex I of H.235 enhances the Baseline Security Profile (Annex D, section 4.2.1.1) as well as the Hybrid Security Profile (Annex F, section 4.2.1.3) with the option to be applied in an environment were direct routed calls (endpoint to endpoint) are performed using the gatekeeper for address resolution. Since endpoints do not possess a shared secret from scratch, a Kerberos-like approach is taken to establish a shared secret between the communicating endpoints. This is done using the admission phase from the calling endpoint and the call signaling between the calling and the called endpoint. The gatekeeper serves in this scenario also as the key distribution center (KDC), issuing two "tickets" (tokens), one containing the key material secured with the caller's encryption key and the other one secured with the called party encryption key. The encryption keys are derived from the shared secret between the caller and the gatekeeper using a pseudo random function (PRF), which is also defined by H.235 Annex I. The PRF is basically the same as used in TLS

The gatekeeper also generates a session key, which is applicable for the communication between the two endpoints involved in the call, and encrypts this key material using the previously derived encryption keys. The encrypted session keys are then transmitted back to the caller. The caller uses the encrypted session key destined to him, the other one is sent to the called party as part of the SETUP message.

The messages exchanged between the gatekeeper and the calling endpoint carrying the tickets are secured with either the H.235 Annex D (section 4.2.1.1) or with H.235 Annex F (4.2.1.4). The shared secret established via the "ticket" (token) exchange between caller and callee may be used in subsequent direct messages to provide an integrity protection according to H.235 Annex D.

4.2.3 H.323 Annex J

H.323 Annex J describes security for simple endpoint types, which are defined by H.323 Annex F. This profile relies on the Baseline Security Profile described in section 4.2.1.1.

4.2.4 H.323 Security Issues

Firewalls pose particularly difficult problems for VOIP networks using H.323. With the exception of the "Q.931-like" H.225, all H.323 traffic is routed through dynamic ports. For H.323 Fast Start and H.245 tunneling just one channel (H.225 Call Signaling) is used. Usually the call signaling is performed via port 1720. If additionally H.225 RAS communication is done with the gatekeeper (UDP), this is done via port 1719. That is, each successive channel in the protocol is routed through a port dynamically determined by its predecessor. This ad-hoc method of securing channels does not lend itself well to a static firewall configuration. This is particularly true in the case of stateless firewalls that cannot comprehend H.323 traffic. These simple packet filters cannot correlate UDP transmissions and replies. This necessitates punching holes in the firewall to allow H.323 traffic to traverse the security bridge on any of the ephemeral ports it might use. This practice would introduce serious security weaknesses because such an implementation would need to leave 10,000 UDP ports and several H.323 specific TCP ports wide open [sample configuration provided in 1]. There is thus a need for a stateful firewall that understands VOIP, specifically H.323. Such a firewall can read H.323 messages and dynamically open the correct ports for each channel as the protocol moves through its call setup process. Such a firewall must be part of a security architecture especially in scenarios where protocol–provided security measures are applied, e.g. message integrity. Barring this, some kind of proxy server or middlebox would have to be used. Several solutions to this problem are presented in chapter 6.

Even with a VOIP-aware firewall, parsing H.323 traffic is not a trivial matter. H.323 traffic is encoded in a binary format based on ASN.1. ASN.1 does not use fixed offsets for address information, and different instances of an application may negotiate different options, resulting in different byte offsets for the same information [21]. This level of complexity does not allow for simple parsing tools or uncomplicated Perl scripts to decode the traffic; in fact special code generators are needed [18]. Such technology is not available on traditional packet filtering firewalls or even simple stateful firewalls. Although this analysis can be done using modern VOIP-aware gateways, the complex parsing necessary to discern the contents of the ASN.1 encoded packets introduces further latency into a speed-sensitive system that is already saturated with delays.

NAT is also particularly troublesome for VOIP systems using the H.323 call setup protocol. NAT complicates H.323 communications because the internal IP address and port specified in the H.323 headers and messages themselves are not the actual address/port numbers used externally by a remote terminal. This disrupts the "setup next" procedure used by each protocol within the H.323 suite (e.g., H.225 setting up H.245). Not only does the firewall have to comprehend this, but it is essential that the VOIP application receiving these H.323 communications receives the correct *translated* address/port numbers. Thus, if H.323 is to traverse a NAT gateway, the NAT device must be able to reconfigure the addresses in the control stream. So with NAT, not only does H.323 traffic need to be read, it must also be modified so that the correct address/port numbers are sent to each of the endpoints.

4.3 Encryption Issues and Performance

Delay in a VOIP system can be added by codecs and by addtional processing such as encryption. Codecs add delay in coding and compressing speech data. Processing time increases with the degree of compression, because larger blocks of speech data are needed to produce higher degrees of compression.

Encryption serves two purposes for VOIP: privacy protection, by encrypting voice data, and message authentication, which protects the origin and integrity of voice packets. Encryption may be done using either a stream or block cipher. If a stream cipher is used, very little delay is introduced if the key stream can be produced before or at least as fast as voice data arrives. In this case there will be only one bit of delay as the cipher stream is applied. Block ciphers may require one block of delay, which will vary with the method used, but still require relatively little overhead.

More significant delays are introduced by computing HMAC hash valures for authentication. HMAC is used with secret key hash functions, such as MD5 or SHA-1. HMAC-MD5 produces a 128-bit message authentication code (MAC), while HMAC-SHA-1 will produce a 160-bit MAC. Because the HMAC operation must wait for a full block of data to arrive before processing, these operations can produce significant latency delays. On arrival, the reverse operations must be applied, introducing further performance delays.

In most applications, authentication and integrity are equally, or more, important than encryption, but with voice processing for human speakers, some authentication is normally built-in because parties recognize the person on the other end of the conversation. Even if the conversation is with a stranger, concern with source authentication applies primarily to call setup, rather than to the authentication of voice packets in the midst of a conversation. As a result of these considerations, some designers may consider HMAC less important for secure VOIP than call encryption, and may limit HMAC use if performance is a problem.

5 SIP

SIP is the IETF specified protocol for initiating a two-way communication session. It is considered by some to be simpler than H.323 [18][16], though it is now the largest RFC in IETF history. SIP is text based; thereby avoiding the ASN.1 associated parsing issues that exist with the H.323 protocol suite, if S/MIME is not used as part of SIP inherent security measures.. Also, SIP is an application level protocol, that is, it is decoupled from the protocol layer it is transported across. It can be carried by TCP, UDP, or SCTP. UDP may be used to decrease overhead and increase speed and efficiency, or TCP may be used if SSL/TLS is incorporated for security services. Newer implementations may use stream control transmission protocol (SCTP), developed in the IETF SIGTRAN working group (RFC 2960) specifically to transport signaling protocols. SCTP offers increased resistance to DoS attacks through a four-way handshake method, the ability to multi-home, and optional bundling of multiple user messages into a single SCTP packet. Additional security services can be used with SCTP via RFC 3436 (TLS over SCTP) or 3554 (SCTP over IP Sec). Unlike H.323, only one port is used in SIP (note that H.323 may also be used in a way that uses only one port – direct routed calls). The default value for this port is 5060.

5.1 SIP Architecture

The architecture of a SIP network is different from the H.323 structure. A SIP network is made up of end points, a proxy and/or redirect server, location server, and registrar. A diagram is provided in Figure 5. In the SIP model, a user is not bound to a specific host (neither is this the case in H.323, gatekeeper provides address resolution). The user initially reports their location to a registrar, which may be integrated into a proxy or redirect server. This information is in turn stored in the external location server.

Messages from endpoints must be routed through either a proxy or redirect server. The proxy server intercepts messages from endpoints or other services, inspects their "To:" field, contacts the location server to resolve the username into an address and forwards the message along to the appropriate end point or another server. Redirect servers perform the same resolution functionality, but the onus is placed on the end points to perform the actual transmission. That is, Redirect servers obtain the actual address of the destination from the location server and return this information to the original sender, which then must send its message directly to this resolved address (similar to H.323 direct routed calls with gatekeeper).

The SIP protocol itself is modeled on the three-way handshake method implemented in TCP (see Figure 6). We will consider the setup here when a proxy server is used to mediate between endpoints. The process is similar with a redirect server, but with the extra step of returning the resolved address to the source endpoint. During the setup process, communication details are negotiated between the endpoints using Session Description Protocol (SDP), which contains fields for the codec used, caller's name, etc. If Bob wishes to place a call to Alice he sends an INVITE request to the proxy server containing SDP info for the session, which is then forwarded to Alice's client by Bob's proxy, possibly via her proxy server. Eventually, assuming Alice wants to talk to Bob, she will send an "OK" message back containing her call preferences in SDP format. Then Bob will respond with an "ACK". SIP provides for the ACK to contain SDP instead of the INVITE, so that an INVITE may be seen without protocol specific information. After the "ACK" is received, the conversation may commence along the RTP / RTCP ports previously agreed upon. Notice that all the traffic was transported through one port in a simple (text) format, without any of the complicated channel / port switching associated with H.323. Still, SIP presents several challenges for firewalls and NAT. These difficulties are discussed in the next section.

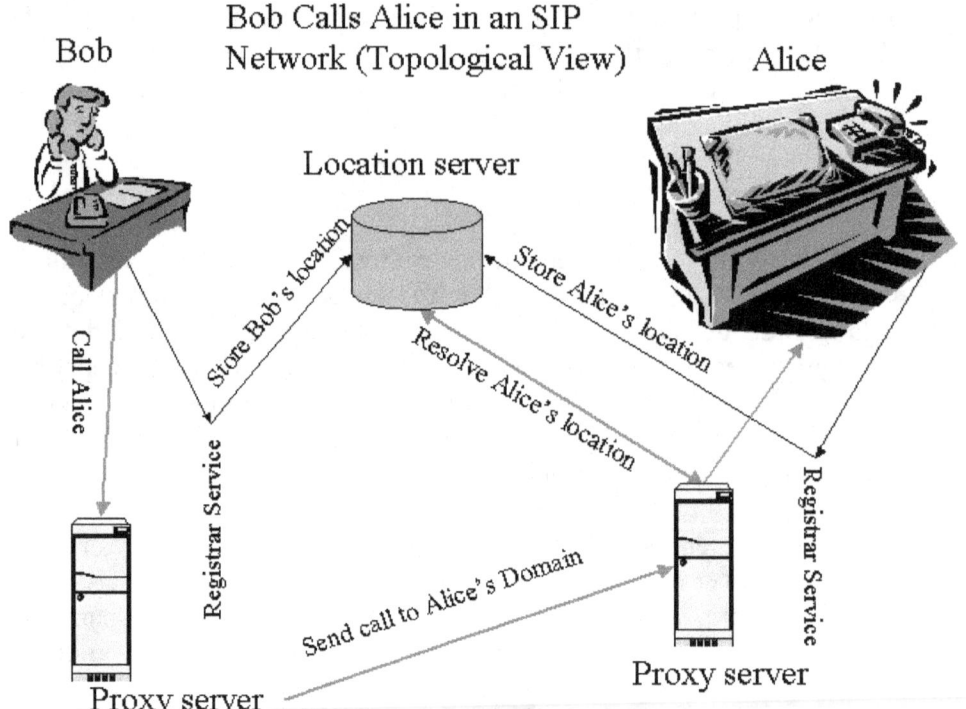

Figure 5. SIP Network Architecture

5.2 Existing Security Features within the SIP Protocol

RFC 3261 describes several security features for SIP, which will be described in the next subsections. RFC 3261 deprecates several security features, which were advocated in the original RFC 2543, such as the usage of PGP and HTTP Basic Authentication.

5.2.1 Authentication of Signaling Data using HTTP Digest Authentication

The Digest authentication scheme is based on a simple challenge-response paradigm. The digest authentication scheme challenges the remote end using a nonce value. SIP digest authentication is based on the digest authentication defined in RFC 2617. Here, a valid response contains a checksum (by default, the MD5 checksum) of the *user name*, the *password*, the given *nonce* value, the *HTTP method*, and the requested *URI*. In this way, the password is never sent in the clear. Because of its weak security, and to avoid attacks by downgrading the required security level of the authentication, HTTP Basic Authentication was not recommended in the current draft of RFC 3261.

5.2.2 S/MIME Usage within SIP

SIP messages carry MIME bodies. MIME itself defines mechanisms for the integrity protection and the encryption of the MIME contents. SIP may use S/MIME to enable mechanisms like public key distribution, authentication and integrity protection, or confidentiality of SIP signaling data. S/MIME may be considered as a replacement for PGP to provide means for integrity protection and encryption of SIP messages. To be able to protect SIP header fields as well, tunneling of SIP messages in MIME bodies is specified. Generally the proposed SIP tunneling for SIP header protection will create additional overhead. S/MIME requires certificates and private keys to be used, whereas the certificates may be issued by a trusted third party or may be self-generated. The latter case may not provide real user authentication but may be used to provide a limited form of message integrity protection. The following sections explain the usage of S/MIME more deeply.

The current document, RFC 3261, recommends S/MIME to be used for UAs. Moreover, if S/MIME is used to tunnel messages (described below) it is recommend using a TCP connection because of the larger messages. This is to avoid problems that may arise by the fragmentation of UDP packets. The following services can be realized: (a) authentication and integrity protection of signaling data, and (b) confidentiality of signaling data

5.2.3 Confidentiality of Media Data

SIP itself does not consider the encryption of media data. Using the RTP encryption as defined in RFC 1889 may provide confidentiality for media data. Another option for media stream security is the use of SRTP [DSRTP]. For key management SDP (cf. RFC 2327) may be used. SDP can convey session keys for media streams. Note that using SDP for the key exchange provides no method to send an encrypted media stream key (cf. Appendix A2). Therefore, the signaling request should be encrypted, preferably by using End-to-End encryption.

5.2.4 TLS usage within SIP

RFC 3261 mandates the use of TLS for proxies, redirect servers, and registrars to protect SIP signaling. Using TLS for UAs is recommended. TLS is able to protect SIP signaling messages against loss of integrity, confidentiality and against replay. It provides integrated key-management with mutual authentication and secure key distribution. TLS is applicable hop-by-hop between UAs/proxies or between proxies. The drawback of TLS in SIP scenarios is the requirement of a reliable transport stack (TCP-based SIP signaling). TLS cannot be applied to UDP-based SIP signaling. Just as secure HTTP is specified with the "https:", secure SIP is specified with a Universal Resource Indicator (URI) that begins with "sips:".

5.2.5 IPsec usage within SIP

IPsec may also be used to provide security for SIP signaling at the network layer. This type of security is most suited to securing SIP hosts in a SIP VPN scenario (SIP user agents/proxies) or between administrative SIP domains. IPsec works for all UDP, TCP and SCTP based SIP signaling. IPsec may be used to provide authentication, integrity and confidentiality for the transmitted data and supports end-to-end as well as hop-by-hop scenarios. At this time there is no default cipher suite for IPsec defined in SIP. Note that RFC 3261 does not describe a framework for the use of IPsec and no requirement is given as to how the key management is to be realized, or which IPsec header and mode is to be used. One accepted protocol for key management is Internet Key Exchange (IKE), a hybrid protocol based on the Internet Security Association and Key Management Protocol (ISAKMP), the Oakley Key Determination Protocol (RFC 2412) and the Secure Key Exchange Mechanism for the Internet (SKEME). The IKE protocol provides automated cryptographic key exchange and management mechanisms for IPsec. IKE is used to negotiate security associations (SAs) for use with its own key management exchanges (called Phase 1) and for other services such as IPsec (called Phase 2). IKE is particularly used in the establishment of VPNs.

5.2.6 Security Enhancements for SIP

Currently within the IETF several drafts concerning security are being discussed, with a view toward providing a general security solution to SIP scenarios. Several drafts have been produced concerning authentication, integrity, and confidentiality for SIP. The following subsections provide a short overview of Internet drafts, which may be of interest for a discussion of security enhancements for common SIP scenarios. This list of Internet drafts is not complete, as this is a continually evolving area, but the most significant drafts are considered here.

5.2.6.1 SIP Authenticated Identity Body

SIP Authenticated Identity Body (AIB) defines a generic SIP authentication token. The token is provided by adding an S/MIME body to a SIP request or response in order to provide reference integrity over its headers. The document defines a format for this message body referred to as an authenticated identity body (AIB). This is a digitally signed SIP message (*sip/message*) or message fragment (*sip/frag*).

5.2.6.2 SIP Authenticated Identity Management

The existing mechanisms for expressing identity in SIP often do not permit an administrative domain to securely verify the identity of the originator of a request. This document recommends practices and conventions for authenticating end users, and proposes a way to distribute cryptographically secure authenticated identities within SIP messages by including an authentication token (as a MIME body). This token is then added to the message.

5.2.6.3 S/MIME AES Requirement for SIP

RFC 3261 specifies 3DES as the required minimum encryption algorithm for implementations of S/MIME in SIP. Although 3DES is still a viable algorithm, NIST has selected an improved algorithm, AES, as a replacement for DES and 3DES. Standards-track RFC 3853, the *S/MIME AES Requirement for SIP*, updates the normative guidance of RFC 3261 to require AES for S/MIME. This specification may be found at: http://www.ietf.org/rfc/rfc3853.txt. AES provides higher throughput and lower computational complexity than 3DES, and can be implemented with low memory requirements, making it more suitable for mobile or embedded devices, including VOIP phones.

5.2.6.4 Security Mechanism Agreement for SIP

SIP has a number of security mechanisms. Some of them have been built in to the SIP protocol directly, such as HTTP authentication. These mechanisms have alternative algorithms and parameters. The idea originates from the 3rd Generation Partnership Project (3GPP), a collaboration of telecommunications companies, and provides a mechanism for selecting which security mechanisms to use between two entities. RFC 3261 itself does not provide any mechanism agreement options. Moreover, even if some mechanisms such as OPTIONS were used to perform a mechanism agreement, the agreement would be vulnerable to Bidding-Down attacks (a phase of man-in-the-middle attack where the attacker modifies messages to convince communicating parties that both sides support only weak algorithms). Three header fields are defined for negotiating the security mechanisms within SIP between a SIP User Agent entity and its next hop SIP server. It is a proposed standard (RFC 3329) from the IETF. Five mechanisms are currently supported:

- TLS
- HTTP Digest
- IPsec with IKE
- manually keyed IPsec without IKE
- S/MIME

5.2.6.5 End-to-Middle, Middle-to-Middle, Middle-to-End Security

Currently there are two drafts being discussed within the IETF dealing with end-to-middle, middle-to-middle and middle-to- end security: "End-to-middle Security in the Session Initiation Protocol (SIP)" [30] and "A Mechanism to Secure SIP information inserted by Intermediaries" [31]. The motivation for the End to Middle Security draft [30] stems from the need to enable intermediaries to use some of the SIP message header and body when end-to-end security is applied. Examples include logging services for enterprise use, firewall traversal, and transcoding (tailoring web pages for varying devices such as PDAs or cell phones). Intermediaries may not be able to trace the SIP message body for certain information (e.g., port numbers to be opened) if the body is encrypted. This draft is related to [31]. There is still a discussion about this draft within the SIPPING group.

The second draft [31] aims at a mechanism to secure information inserted by intermediaries. This document came about to provide a more robust security solution for History-Info header, however, the intention is to provide an overall more robust security solution for SIP. Proxies sometimes have the need to read and/or modify a message body or header in a request. However, this is explicitly precluded by the SIP specification and is further complicated when a message body is protected with S/MIME. RFC 3261 is designed so that a proxy does not break integrity of the body.

The security requirements between both approaches are slightly different, since information is added by intermediaries and used by intermediaries. Nevertheless, SIP End to Middle Security [30] and SIP Intermediate Security [31] share the same fundamental problems to be solved in SIP. It is anticipated that there will be further discussion on this item, as certain scenarios exist where this functionality is needed.

5.2.7 SIP Security Issues

The text encoding of SIP makes it easier to analyze using standard parsing tools such as Perl or lex and yacc. Still, some new requirements are placed on the firewall in a SIP-based VOIP network. First, firewalls must be stateful and monitor SIP traffic to determine which RTP ports are to be opened and made available to which addresses. This responsibility is similar to the task firewalls on an H.323 based network perform, except the call setup and header parsing is much simpler. The other issues SIP-based VOIP encounters with firewalls are associated with RTP traffic and incoming calls, as covered in section 7.4. As with H.323, the big problem for SIP is NAT.

NAT inhibits SIP's registration and communication mechanisms and requires innovative solutions to resolve these issues. The problems exist because in a SIP-based network, the SIP proxy is normally outside the NAT device. There are three main scenarios for using a SIP proxy:
- The proxy is within the corporate LAN and the Teleworker connects from outside
- The proxy is at the telecom side and clients from, for instance, smaller companies connecting to this proxy for VOIP service
- Two administrative domains are connected, both have their own proxy.

So the problem is bartering communication between a proxy server that deals with global IP addresses and a machine that has been assigned a private network address. Rosenberg & Schulzrinne [22] classify three different sets of problems SIP traffic has in such an architecture: originating requests, receiving requests, and handling RTP. We have already dealt with the incompatibilities of RTP with NAT and now we will see the issues NAT presents to the call setup process itself.

To initialize a session from behind the NAT, a caller can simply send an INVITE message as always. The outgoing port number (5060) will be preserved by the NAT, but response communication could be disturbed. If SIP is implemented over UDP (recall SIP is protocol independent) the proxy server must send the UDP response to the address and port the request arrived on [22] A simpler solution is to use the standard practice of routing SIP communication over TCP. With TCP, the response from the callee will come over the same channel as the original INVITE and so NAT will not present a problem.

Bob Calls Alice in an SIP Network (Protocol View)

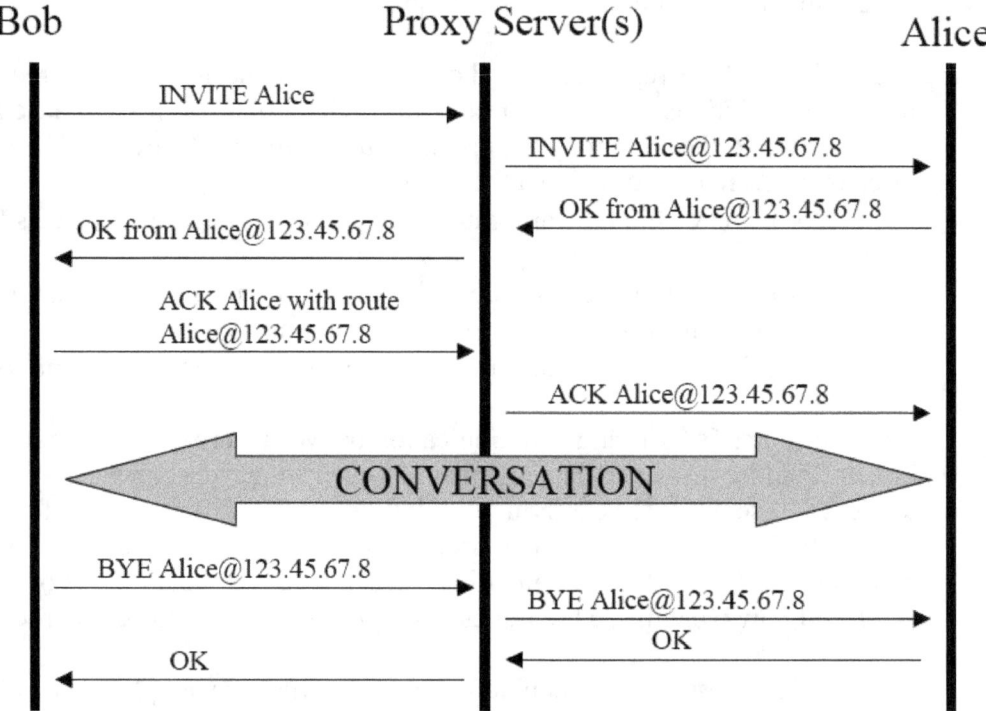

Figure 6. SIP Protocol

We have already discussed some of the problems with incoming VOIP connections against NAT (see Section 4.2). Now we will look more in depth at the SIP specific problems with incoming calls. Rosenberg & Schulzrinne [22] trace the problem back to the registration process itself. When a user contacts the registrar, they provide their IP address as their reachable address and this is stored in the location server. However, this is their private IP address. The proxy server deals only with global IP addresses, so when a message comes in for username@domain.com, it will attempt to route this call to the registered address, but in the public domain. For instance, if username@domain.com is registered to an internal IP address of 10.7.34.189, then the proxy server will attempt to forward the traffic to this address, but in the public domain. This address is unreachable for the proxy server and the connection will be refused. The solution to this is a delicate manipulation of IP addresses and an expansion of the responsibilities of the SIP proxy server.

6 Gateway Decomposition

Media gateway control protocols address the requirements of IP telephony networks that are built using "decomposed" VOIP gateways. Decomposed VOIP gateways consist of Media Gateways (MGs) and Media Gateway Controllers (MGC), and appear to the outside as a single VOIP gateway. MGC handles the signaling data between the MGs and other network components such as H.323 gatekeepers or SIP Servers, or towards SS7 Signaling Gateways. MGs focus on the audio signal translation function, performing conversion between the audio signals carried on telephone circuits and data packets carried over the Internet or other packet networks. A single MGC can control multiple MGs, which leads to cost reductions when deploying larger systems. Common examples are the Media Gateway Control Protocol (MGCP) and Megaco/H.248, which are described within the next subsections.

6.1 MGCP

6.1.1 Overview

MGCP is used to communicate between the separate components of a decomposed VOIP gateway. It is a complementary protocol to SIP and H.323. The MGCP protocol was derived from version 1.1 of the SGCP protocol, which was a fusion of the SGCP version 1 and IPDC. MGCP is currently being maintained by PacketCable (called NCS (Network Call Signaling Protocol)) and the Softswitch Consortium. In October 1999, MGCP was converted into an informational RFC 2705.

There are plans for the MGCP specification to be enhanced by international standards bodies. One is the IP Cablecom activity proposing J.162 (Network Call Signaling) and J.171 (Trunking Gateway Control Protocol, a variant of J.162). A similar version of these proposals will also be provided within ETSI as EuroPacketCable specifications. At the present time, MGCP is the de-facto industry standard and has not yet been superseded by MEGACO/H.248.

6.1.2 System Architecture

Within MGCP the MGC server or "call agent" is mandatory and manages calls and conferences, and supports the services provided (see Figure 7). The MG endpoint is unaware of the calls and conferences and does not maintain call states. MGs are expected to execute commands sent by the MGC call agents. MGCP assumes that call agents will synchronize with each other sending coherent commands to MGs under their control. MGCP does not define a mechanism for synchronizing call agents. MGCP is a master/slave protocol with a tight coupling between the MG (endpoint) and MGC (server).

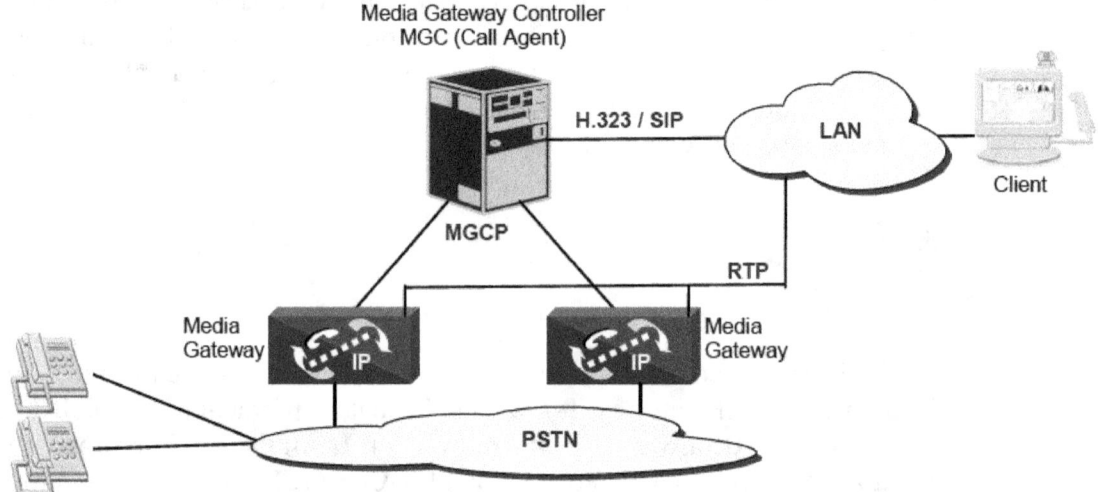

Figure 7: General Scenario for MGCP Usage

RTP data is exchanged directly between the involved media gateways. The Call Agent uses MGCP to provide the gateways with the description of connection parameters such as IP addresses, UDP port and RTP profiles. These descriptions follow the conventions delineated in the Session Description Protocol (SDP) from RFC 2327. SDP is a session description protocol for multimedia sessions, which runs also on UDP connections.

6.1.3 Security Considerations

There are no security mechanisms designed into the MGCP protocol itself. The informational RFC 2705 refers to the use of IPsec (either AH or ESP) to protect MGCP messages. Without this protection a potential attacker could set up unauthorized calls or interfere with ongoing authorized calls. Beside the use of IPsec, MGCP allows the call agent to provide gateways with session keys that can be used to encrypt the audio messages, protecting against eavesdropping. The session key will be used later on in RTP encryption. The RTP encryption, described in RFC 1889, may be applied. Session keys may be transferred between the call agent and the gateway by using the SDP (cf. RFC 2327).

6.2 Megaco/H.248

6.2.1 Overview

The IETF started to work on MEGACO as a compromise protocol between MGCP and MDCP. ITU-T SG16 adopted MEGACO version 0.1 in April 1999 as the starting specification for H.GCP (H-series, Gateway Control Protocol), later H.248. In June 1999, the IETF MEGACO WG and ITU-T came out with a single document describing a standard protocol for interfacing between Media Gateway Controllers (MGCs) and Media Gateways (MGs) MEGACO/H.248. MEGACO/H.248. It is expected to win wide industry acceptance as the official standard for decomposed gateway architectures released by both the IETF and ITU-T.

Since MEGACO/H.248 is derived from MGCP, many similarities can be found, for instance:

- Similarity between the semantics of the commands in the two specifications.
- The use of ABNF grammar for syntax specification and the Session Description Protocol (SDP) to specify media stream properties is the same as in MGCP.
- The processing of signals and events in media streams is the same in MEGACO as in MGCP.
- The concept of packages containing event and signal definitions that permits easy extension to the protocol is borrowed from MGCP.
- The MEGACO specification for transport of messages over UDP is the same as specified in MGCP. The three-way-handshake and the computation of retransmission timers described in MGCP are also described within the ALF definition specified in Annex E of MEGACO.

MEGACO/H.248 introduces several enhancements compared with MGCP, including the following:

- Support of multimedia and multipoint conferencing enhanced services
- Improved syntax for more efficient semantic message processing
- TCP and UDP transport options
- Allows either text or binary encoding (to support IETF and ITU-T approach)
- Formalized extension process for enhanced functionality
- Expanded definition of packages

MEGACO is described as Gateway Control Protocol Version 1 within the RFC 3525.

6.2.2 System Architecture

MEGACO/H.248 (see Figure 8) has basically the same architecture as MGCP. MEGACO/H.248 commands are similar to MGCP commands. However, the protocol models are quite different. MEGACO specifies a media gateway connection model that has two entities: Terminations (source or sink for (one or more) media streams), and context (grouping of terminations connected in a call). In contrast, MGCP uses the following two entities: Endpoints (source or sink of data), and Connection: (association between two endpoints).

Taking a multipoint conference as an example, MEGACO simplifies the connection setup by adding terminations to a context, whereas MGCP has to establish several connections to the conference server. The context in this scenario may cover multiple media streams for enhanced multimedia services.

With MEGACO/H.248, the primary mechanism for extension is by means of packages. In general, MEGACO/H.248 Packages include more detail than MGCP Packages. They define additional properties and statistics along with event and signal information that may occur on terminations.

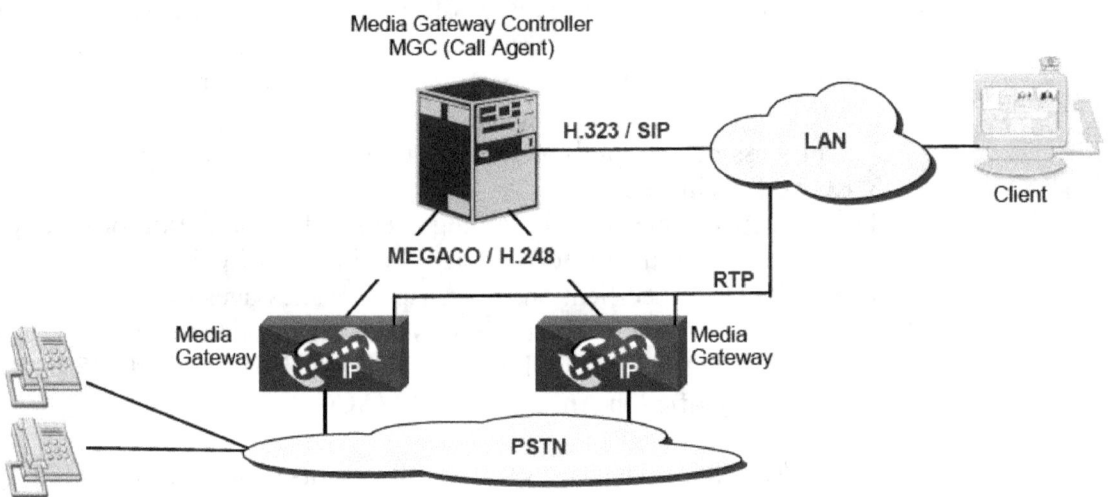

Figure 8: General Scenario for MEGACO/H.248 Usage

6.2.3 Security Considerations

Megaco (RFC 3525) recommends security mechanisms that may be in underlying transport mechanisms, such as IPsec. H.248 goes a step further by requiring that implementations of the H.248 protocol implement IPsec if the underlying operating system and the transport network support IPsec. Implementations of the protocol using IPv4 are required to implement the interim AH scheme. H.248 states that implementations employing the AH header *shall* provide a minimum set of algorithms for integrity checking using manual keys (compliant to RFC 2402).

The interim AH scheme is the use of an optional AH header, which is defined in the H.248 protocol header. The header fields are exactly those of the SPI,

SEQUENCE NUMBER and DATA fields as defined in RFC 2402. The semantics of the header fields are the same as the "transport mode" of RFC 2402, except for the calculation of the Integrity Check Value. For more details on the calculation of the ICV check H.248. The interim AH interim scheme does not provide protection against the eavesdropping and replay attacks.

For MEGACO, manual key management is assumed and replay protection, defined for IPsec, may not be used in this scenario (the sequence number in the AH may overrun when using manual key management, since re-keying is not possible). Furthermore, H.248 states that implementations employing the ESP header *shall* provide a minimum set of algorithms for integrity checking and encryption (compliant to RFC 2402). Moreover, implementations *should* use IKE (RFC 2409) to permit more robust keying options. Implementations employing IKE should support authentication with RSA signatures and RSA public key encryption.

7 Firewalls, Address Translation, and Call Establishment

Firewalls and NAT present a formidable challenge to VOIP implementers. However, there are solutions to these problems, if one is willing to pay the price. Commonly used solutions that can be integrated into a standard network configuration containing a firewall and/or NAT are presented here. Much of the standards work thus far has been done for SIP in the IETF, though the ITU has recently taken up the firewall/NAT traversal issue for H.323 in Study Group 16. It is important to note that all three major VOIP protocols, SIP, H.323, and H.248/MEGACO all have similar problems with firewalls and NATs. Although the use of NATs may be reduced as IPv6 is adopted, they will remain a common component in networks for years to come, and IPv6 will not alleviate the need for firewalls, so VOIP systems must deal with the complexities of firewalls and NATs.

7.1 Firewalls

Firewalls are a staple of security in today's IP networks. Whether protecting a LAN, WAN, encapsulating a DMZ, or just protecting a single computer, a firewall is usually the first line of defense against would be attackers. Firewalls work by blocking traffic deemed to be invasive, intrusive, or just plain malicious from flowing through them. If networks are castles, firewalls are the drawbridges. Traffic not meeting the requirements of the firewall is dropped. Processing of traffic is determined by a set of rules programmed into the firewall by the network administrator. These may include such commands as "Block all FTP traffic (port 21)" or "Allow all HTTP traffic (port 80)". Much more complex rule sets are available in almost all firewalls.

A useful property of a firewall, in this context, is that it provides a central location for deploying security policies. It is the ultimate bottleneck for network traffic because when properly designed, no traffic can enter or exit the LAN without passing through the firewall. This situation lends itself to the VOIP network where firewalls simplify security management by consolidating security measures at the firewall gateway, instead of requiring all the endpoints to maintain up to date security policies. This takes an enormous burden off the VOIP network infrastructure. Note that this abstraction and simplification of security measures comes at a price. The introduction of firewalls to the VOIP network complicates several aspects of VOIP, most notably dynamic port trafficking and call setup procedures. This chapter describes various complications firewalls introduce into the system. But since firewalls are an essential and often already deployed component of the modern network, we will also examine some proposed and applied solutions to these entanglements.

7.1.1 Stateful Firewalls

Most VOIP traffic travels across UDP ports. Firewalls typically process such traffic using a technique called packet filtering. Packet filtering investigates the headers of each packet attempting to cross the firewall and uses the IP addresses, port numbers, and protocol type (collectively known as the 5-tuple) contained therein to determine the packets' legitimacy. In VOIP and other media streaming protocols, this information can also be used to distinguish between the start of a connection and an established connection. There are two types of packet filtering firewalls, stateless and stateful. Stateless firewalls retain no memory of traffic that has occurred earlier in the session. Stateful firewalls do remember previous traffic and can also investigate the application data in a packet. Thus, stateful firewalls can handle application traffic that may not be destined for a static port.

7.1.2 VOIP specific Firewall Needs

In addition to the standard firewall practices, firewalls are often deployed in VOIP networks with the added responsibility of brokering the data flow between the voice and data segments of the network. This is a crucial functionality for a network containing PC-Based IP phones that are on the data network, but need to send voice messages. All voice traffic emanating from or traveling to such devices would have to be explicitly allowed in if no firewall was present because RTP makes use of dynamic UDP ports (of which there are thousands). Leaving this many UDP ports open is an egregious breach of security. Thus, it is recommended that all PC-based phones be placed behind a stateful firewall to broker VOIP media traffic. Without such a mechanism, a UDP DoS attack could compromise the network by exploiting the plethora of open ports. This is one example of how firewalls are used to provide a logical segmentation of the network, providing a barrier between voice and data sectors. Halpern [23] identifies some of the key points of collision between voice and data traffic where firewalls are necessary, including:

- PC-Based IP phones (data) require access to the (voice) segment to place calls, leave messages, etc.
- IP Phones and call managers (voice) accessing voice mail (data),
- users (data) accessing the proxy server (voice)
- the proxy server (voice) accessing network resources (data).
- traffic from IP Phones (voice) to the call processing manager (voice) or proxy server (voice) must pass through the firewall because such contacts use the data segment as an intermediary.

Halpern [23] provides details for all these connections. Firewalls could also be used to broker traffic between physically segmented traffic (one network for VOIP, one network for data) but such an implementation is fiscally and physically unacceptable for most organizations, since one of the benefits of VOIP is voice and data sharing the same physical network.

7.2 Network Address Translation

Network Address Translation (NAT) is a powerful tool that can be used to hide internal network addresses and enable several endpoints within a LAN to share the same (external) IP address. For the purposes of this document, NAT actually refers to Network Address and Port Translation (NAPT). In NAT as it is literally defined, outgoing IP headers are changed from private LAN addresses to the router's global IP. In NAPT, the TCP/UDP headers themselves are converted. This allows several computers to simultaneously share the router's global IP address. Also, machines that do not need to access the Internet can still be assigned local addresses on the intranet without producing conflicts or needlessly taking up an IP address. With the shortages of IP addresses in many regions, this is an extremely useful functionality.

NATs also indirectly contribute to security for a LAN, making internal IP addresses less accessible from the public Internet. Thus, all attacks against the network must be focused at the NAT router itself. Like firewalls, this provides security because only one point of access must be protected, and the router will generally be far more secure than a PC directly connected to the Internet (less likelihood of open ports, malicious programs, etc.). The abstraction of the LAN from the Internet through a NAT also simplifies network management. For instance, if one decided to change their ISP, only the external router configuration would need to be changed. The internal network and addressing scheme could be left untouched [24].

Different types of NAT policies result in different complexity (the following terminology is adopted from the STUN RFC [32] describing a NAT traversal methodology, which originates from the MIDCOM working group of the IETF):

Full Cone NAT
A full cone NAT is one where all requests from the same internal IP address and port are mapped to the same external IP address and port. Furthermore, any external host can send a packet to the internal host, by sending a packet to the mapped external address.

Restricted Cone NAT
Restricted Cone: A restricted cone NAT is one where all requests from the same internal IP address and port are mapped to the same external IP address and port. Unlike a full cone NAT, an external host (with IP address X) can send a packet to the internal host only if the internal host had previously sent a packet to IP address X.

Port Restricted Cone
> A port restricted cone NAT is like a restricted cone NAT, but the restriction includes port numbers. Specifically, an external host can send a packet, with source IP address X and source port P, to the internal host only if the internal host had previously sent a packet to IP address X and port P. It is used to enable sharing of external IP addresses.

Symmetric NAT
> A symmetric NAT is one where all requests from the same internal IP address and port, to a specific destination IP address and port, are mapped to the same external IP address and port. If the same host sends a packet with the same source address and port, but to a different destination, a different mapping is used. Furthermore, only the external host that receives a packet can send a UDP packet back to the internal host.

All of these benefits of NAT come at a price. NATs "violate the fundamental semantic of the IP address, that it is a globally reachable point for communications" [25]. This design has significant implications for VOIP. For one thing, an attempt to make a call into the network becomes very complex when a NAT is introduced. The situation is analogous to a phone network where several phones have the same phone number, such as in a house with multiple phones on one line (see
Figure 9). There are also several issues associated with the transmission of the media itself across the NAT, including an incompatibility with IPsec detailed in section 8.4.

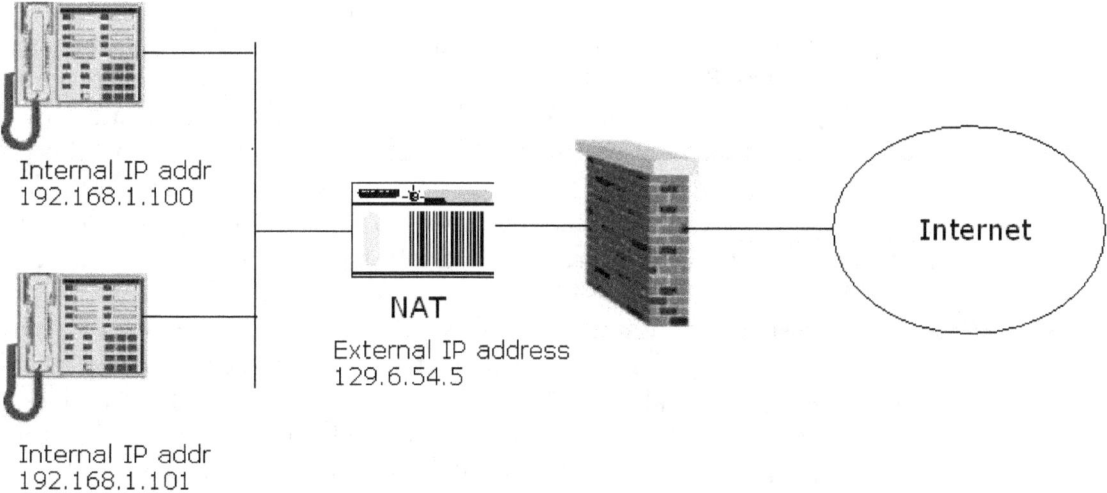

Figure 9. IP Telephones Behind NAT and Firewall

Conceptually, the easiest solution to these incompatibilities is to do away with

NATs entirely, but NATs have their benefits, and even if IPv6 and its expanded address space were implemented today and enough IP addresses were available for everyone to have their own unique IP's, there would still be a need for NATs. Some ISPs use a scheme where users are assigned static IP addresses, one per user. It is unlikely that an ISP would completely overhaul its system and move to a dynamic IP assignment (i.e. DHCP) just because a wealth of new addresses are available to IPv6. This would undermine their whole network and lead to vulnerabilities and opportunities for malicious users to steal Internet access. But many users will still want to connect multiple machines to the Internet using a single IP address, and so the use of NATs will continue. There are many scenarios analogous to this one where NATs are both the cheapest, easiest, and most efficient solution, so NATs are not likely to be abandoned.

7.3 Firewalls, NATs, and VOIP Issues

Some VOIP issues with firewalls and NATs are unrelated to the call setup protocol used. Both network devices make it difficult for incoming calls to be received by a terminal behind the firewall / NAT. Also, both devices affect QoS and can wreak havoc with the RTP stream. The following sections describe these non-protocol specific issues.

7.3.1 Incoming Calls

Regardless of the protocol used for call setup, firewalls and NATs present considerable difficulties for incoming calls. Allowing signal traffic through a firewall from an incoming call means leaving several ports open that might be exploited by attackers. Careful administration and rule definitions should be used if holes are to be punched in the firewall allowing incoming connections. Solutions exist without such holes, including Application Level Gateways and Firewall Control Proxies. NAT creates even more difficulties for incoming calls. Any IP application, including VOIP, that needs to make a connection from an external realm to a point behind a NAT router, would need to know this point's external IP and port number assigned by the router [10]. This situation is far from ideal because it precludes a caller outside the NAT from reaching an internal address except in extreme circumstances. In fact, with dynamic ports being assigned by the NAT, this is nearly an impossible situation because the port the caller requests will be changed by the NAT. Thus, an IP telephony endpoint behind a NAT being analogous to a phone behind a switchboard such that it can only make outgoing calls. For endpoints behind firewalls and NATs, it may be necessary to publish the contact address to enable other clients to call them. This is not an acceptable solution for thousands of people using NAT today. However, there are some solutions to this problem (see chapter 8).

7.3.2 Effects on QoS

Both firewalls and NATs can degrade QoS in a VOIP system by introducing latency and jitter. NATs can also act as a bottleneck on the network because all traffic is routed through a single node. QoS related issues stemming from these network staples are presented here.

VOIP is highly sensitive to latency. So a firewall needs to be able to broker data traffic, but cannot incur time penalties of any significant length, even those that would go unnoticed with simple data traffic [26]. At issue is not only how fast the firewall can interact with the network traffic, but how fast its processor can handle VOIP packets. Two aspects of VOIP can cause degraded behavior in the firewall CPU. First, the call setup process has to be done using H.323 or SIP (see chapters 4-5). Regardless of the protocol used, firewalls need to "dig deeper" into these packets to determine their validity. A flood of call request packets, as the result of an increase in call volume or a malicious attack, can intensify this effect. The presence of a NAT compounds this issue because the payload of the packet must then be changed at the application level to correspond to the NAT translated source or destination address and ports, requiring not only "digging" but filling in the hole with new dirt as well. All this labor puts a tremendous burden on the firewall processor, which must accomplish all these tasks while introducing the bare minimum in latency, especially if protocol security measures are used, such as message integrity.

The other aspect of VOIP that can put a strain on a firewall CPU is the small but plentiful number of RTP packets that make up a VOIP conversation. Firewalls are rarely concerned with the size of a packet, but since each packet must be inspected, a large number of packets can stress the firewall. For example, a firewall may support 100 Mbit/sec (based on the assumption of large packets), but may be overloaded by a flood of small 50 byte packets long before the 100 Mbit/sec rate is reached [LaCour qtd in 26]. QoS/VOIP aware firewalls are designed to avoid performance problems such as these.

No matter how fast the network connection, the firewall CPU is a bottleneck for all unencrypted network packets. Thus a solution to this issue is to use a VPN for all VOIP traffic. Discussion of the pros and cons of this choice is presented in chapter 8.

7.3.3 Firewalls and NATs

Firewalls have difficulties sorting through VOIP signaling traffic. There are solutions to this but there is an additional, and even more vexing problem associated with firewalls and VOIP media. RTP traffic is dynamically assigned an even port number in the range of UDP ports (1024-65534). In addition, the RTCP port controlling this stream will flow through an associated, randomly-assigned port. Allowing such traffic along such a vast number of ports by default would leave the system highly exposed. So firewalls must be made aware dynamically of which ports media is flowing across and between which terminals. For this reason, only stateful firewalls that can process H.323 and SIP should be incorporated into the network to open and close ports. Many new firewalls come equipped with such functionality, although sometimes they support only one protocol (H.323 or SIP) [see 27 for a partial list]. If such firewalls are not available/feasible there are additional hardware solutions available, or VPNs can be used to tunnel through the firewall (chapter 8)

NATs also introduce major design complications into media traffic control in VOIP. First of all, the standard NAT practice of assigning new port numbers at random breaks the pair relationship of RTP and RTCP port numbers [7]. The translation of IP Addresses and ports by NAT is also problematic for the reception of VOIP packets. If the NAT router does not properly process the traffic, the new addresses/ports will not correspond to those negotiated in the call setup process. In this scenario, the VOIP gateway may not properly deliver the RTP packets. The problem is exacerbated if both call participants are behind NATs. Several solutions are available for users with administrative control of their network security and the money to upgrade their firewall/NAT hardware (see 7.1-7.2). For users without control of their network architecture, there is a proposed solution [25], which is detailed in section 9.6. However, this solution breaks down when both users are behind NATs and it has not been implemented successfully to date.

The use of NATs adds another possible complication to VOIP call signaling due to the finite nature of NAT bindings. At a NAT, a public IP address is bound to a private one for a certain period of time (t). This entry gets deleted if no traffic was observed at the NAT for t seconds or the connection was torn down explicitly. A SIP INVITE message, which triggers the binding of the private address to the public one, establishes "state" information when it passes through NATs or firewalls. This state eventually must be cleared. When TCP is used, closure of the TCP connection is usually a good indicator of the termination of the application. However, when SIP runs over UDP such a clear indication is missing. Furthermore, as a silence period during a conversation might be longer than t seconds, not receiving traffic for t seconds might not suffice as an indication of session termination. As a result, it is possible that some state information is destroyed before the transaction and/or the call actually completes.

7.4 Call Setup Considerations with NATs and Firewalls

VOIP users will not tolerate excessive latency in the call setup process, which corresponds to lifting the receiver and dialing in a traditional system. Users may be annoyed with a setup process that requires more than a few seconds [12]. Many factors influence the setup time of a VOIP call. At the network level, these include the topology of the network and the location of both endpoints as well as the presence of a firewall or NAT. At the application level, the degree or lack of authentication and other data security measures, as well as the choice of protocol used to set up the call, can dramatically alter the time necessary to prepare a VOIP connection.

7.4.1 Application Level Gateways

Application Level Gateways (ALGs) are the typical commercial solution to the firewall/NAT traversal problem [10]. An ALG is embedded software on a firewall or NAT, that allows for dynamic configuration based on application specific information. A firewall with a VOIP ALG can parse and understand H.323 or SIP, and dynamically open and close the necessary ports. When NAT is employed, the ALG needs to open up the VOIP packets and reconfigure the header information therein to correspond to the correct internal IP addresses on the private network, or on the public network for outgoing traffic. This includes modifying the headers and message bodies (e.g., SDP) in H.323 and SIP. ALG implementations are discussed for H.323 in [21] and SIP in [22]. The NAT problem is alleviated when the ALG replaces the private network addresses with the address of the ALG itself. It works by not only changing the IP address, but also mapping RTP traffic into ports the ALG can read from and forward to the correct internal machine. The need for consecutive ports for RTP and RTCP can cause a problem here [22] because all VOIP traffic on the network (and data traffic as well) is being routed through the ALG, so as call volume increases, finding enough consecutive ports may become an issue. So although both endpoints may have adequate ports to convene a conversation, the firewall's deficiencies may cause the call to be rejected as "busy" by the ALG itself.

There are significant performance and fiscal costs associated with the implementation of an ALG. Performance-wise, the manipulation of VOIP packets introduces latency into the system and can contribute to jitter when high call volumes are experienced. Depending on the firewall architecture, this can also slow down throughput in the firewall, contributing to general network congestion. A firewall with ALG support can be expensive, and would need to be upgraded or replaced each time the standards for VOIP change. Also, the addition of application intelligence to a firewall can introduce instabilities into the firewall itself. Some firewalls have been found vulnerable to an attack in which a high rate of call setups can be sent, depleting the connection tables of the firewall. These half-open VOIP sessions may not time out in the firewall for more than 24 hours. Still with all these detractions, an ALG remains the simplest and safest workaround to allow the coexistence of VOIP, firewalls, and NAT.

7.4.2 Middlebox Solutions

One drawback to ALGs is that they are embedded in the firewall itself, and thus the latency and throughput slowdown of all traffic traversing the firewall is aggregated and then compounded by the VOIP call volume. Middlebox-style solutions attempt to alleviate this malady by placing an extra device outside the firewall that performs many of the functions associated with an ALG. The device that the application intelligence is extracted to can be an "in-path" system such as an H.323 gatekeeper or a SIP Proxy that sits in the control path of the session and is considered to be a "trusted system" [28] that parses VOIP traffic and instructs the firewall to open or close ports based on the needs of the VOIP signaling via a midcom protocol (see Figure 10). The midcom protocol has not been finalized yet by the IETF. Abstracting stateful inspection and manipulation of signaling packets from the NATs and firewalls (middleboxes) will improve scalability and in the long run, reduce the cost of updating the network[10] by not having to replace the firewall every time the protocols change. There is also a performance improvement that comes from abstracting two highly processor intensive tasks (VOIP parsing and packet filtering) into two separate spheres of influence. This strategy is currently being pursued by the IETF in the Middlebox Communications (Midcom) Working Group.

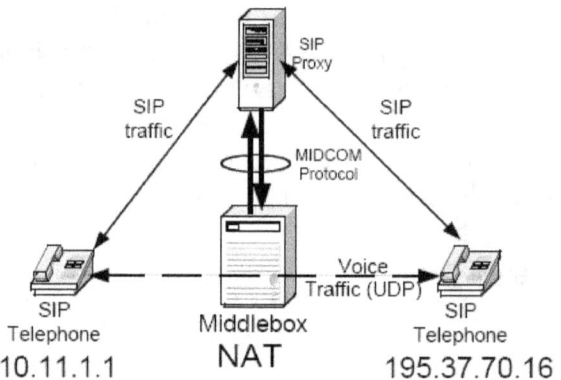

Figure 10. Middlebox Communications Scenario

There are some drawbacks to this approach. First, the firewall must be configured for control by the application-aware device, which incurs an initial setup cost. Also, the middlebox itself requires protection from attackers. A compromised midcom agent is disastrous for the network at large because the firewall takes control cues from the "trusted" device running the midcom agent. Thus an intruder taking control of the midcom agent could open any ports in the firewall and then gain access to the private network. So if the application aware device (like a SIP Proxy) is placed outside the firewall, a second firewall would have to be used to protect that device.

7.4.3   *Session Border Controllers*

While application level gateways may carry scalability concerns, middlebox solutions have not found their way out of standards bodies and into commercial products as fast as might have been hoped. In the absence of a universally accepted solution to the issues associated with firewall/NAT traversal, product developers have brought to market a solution that has come to be known as a

Session Controller, or a Session Border Controller (SBC). SBCs are dedicated appliances that offer one or more of the following services to a VOIP perimeter: Firewall/NAT traversal, Call Admission Control, Service Level Agreement monitoring, support for lawful intercept, and protocol interworking. Third party analysis of these solutions is not widely available as of yet, but in the near term, the demand for these products is expected to grow.

7.5 Mechanisms to solve the NAT problem

Especially for real-time communication protocols like H.323 and SIP, NAT causes trouble because these protocols include IP addresses in their messages. For integrity protection the NAT device would have to be a trusted intermediate host to recalculate the integrity checksum. From an end-to-end security point of view this is not recommended. Therefore the following section describes mechanisms to handle the NAT problem differently.

Simple Traversal of UDP through NATs (STUN)

As defined in RFC 3489, Simple Traversal of User Datagram Protocol (UDP) Through Network Address Translators (NATs) (STUN) is a lightweight protocol that allows applications to discover the presence and types of NATs and firewalls between them and the public Internet. It also provides the ability for applications to determine the public Internet Protocol (IP) addresses allocated to them by the NAT (address bindings). STUN works with many existing NATs, and requires no changes to NATs. STUN does not work with symmetric NAT, because the IP address port mapping is dependent on the destination, i.e. The STUN server would deliver the IP address port mapping for the connection to the STUN server itself, which is different from the mapping for the destination client.

Traversal Using Relay NAT (TURN)

The draft for Traversal Using Relay NAT (TURN) was submitted by J.Rosenberg (dynamicsoft), R. Mahy (Cisco), C. Huitema (Microsoft) in October 2003 and revised July 19, 2004. TURN is a protocol that allows for an element behind a NAT or firewall to receive incoming data over TCP or UDP connections to complement the limitations of STUN. The connection has to be requested by the TURN client, TURN supports the connection of a user behind a NAT to only a single peer. The TURN server would act as a data relay, receiving data on the address it provides to clients, and forwarding them to the clients. TURN is identical in syntax and general operation to STUN, but allocates transport address bindings.

Unlike a STUN server, a TURN server provides resources (bandwidth and ports) to clients that connect to it. Therefore, only authorized clients can access the TURN server. TURN assumes the existence of a long-lived shared secret between the client and the TURN server in order to achieve authentication of the TURN requests. The client uses this shared secret to authenticate itself in a Shared Secret

Request, sent over TLS. The Shared Secret Response provides the client with a one-time username and password. This password is then used to authenticate the Allocate message sent by the client to the TURN server to ask for a public IP address and port. For incoming data, the TURN server then stores the remote address and port where the data came from and forwards the data to the client. The TURN server is responsible for guaranteeing that packets sent to the public IP address route to the TURN server. TURN also allows a client to request an odd or even port when one is allocated, and for it to pre-allocate the next higher port as is useful, e.g. for the H.323 or RTP protocol.

Interactive Connectivity Establishment (ICE)

Interactive Connectivity Establishment (ICE) is an IETF draft, that has also been submitted by J.Rosenberg (dynamicsoft), R. Mahy (Cisco), C. Huitema (Microsoft). ICE describes a methodology for NAT-Traversal for the SIP protocol. ICE is not a new protocol, but makes use of existing protocols, such as STUN, TURN and Real Specific IP (RSIP). ICE works through the mutual cooperation of both endpoints in a SIP dialog. ICE does not require extensions from STUN, TURN or RSIP. However, it does require some additional SDP attributes.

Universal Plug and Play (UPnP)

Another solution originally designed for the home market is Universal Plug and Play. In this scenario, the NAT is upgraded to support the UPnP protocol, and the client can query the NAT directly as to its external IP Address and Port number. However, UPnP does not scale to cascaded NATs, and there are potentially serious security issues with this solution, including vulnerability to denial of service attacks. For more information, refer to www.upnp.org

7.6 Virtual Private Networks and Firewalls

The next chapter deals with the integration of VPN tunneling, and specifically IPsec tunneling into the VOIP system. VPNs alleviate many of the problem issues set forth in this chapter by tunneling straight through firewalls. Hence, many of the firewall specific issues mentioned here become moot. However, this "inelegant" method has some drawbacks. First, tunneling all VOIP traffic over VPNs prohibits firewalls from investigating incoming and outgoing packets for malicious traffic. Also, the centralization of security at the firewall is virtually lost. Also, VPN tunneling with IPsec can be incompatible with NAT. Finally, a host of new QoS and security issues are introduced by the integration of IPsec into VOIP. The next chapter covers these issues in greater detail.

8 Encryption & IPsec

Thus far we have focused primarily on security of the network, protecting endpoints, gateways, and other components, from malicious attacks. Firewalls, gateways, and other such devices can help keep intruders from compromising a network, but firewalls are no defense against an internal hacker. Another layer of defense is necessary at the protocol level to protect the data itself. In VOIP, as in data networks, this can be accomplished by encrypting the packets at the IP level using IPsec. This way if anyone on the network, authorized or not, intercepts VOIP traffic not intended for them (for instance via a packet sniffer), these packets will be unintelligible. The IPsec suite of security protocols and encryption algorithms is the standard method for securing packets against unauthorized viewers over data networks and will be supported by the protocol stack in IPv6. Hence, it is both logical and practical to extend IPsec to VOIP, encrypting the signal and voice packets on one end and decrypting them only when needed by their intended recipient. But the nature of the signaling protocols and the VOIP network itself prevent such a simple scheme from being used, as it becomes necessary for routers, proxies, etc. to read the VOIP packets. Also, several factors, including the expansion of packet size, ciphering latency, and a lack of QoS urgency in the cryptographic engine itself can cause an excessive amount of latency in the VOIP packet delivery. This leads to degraded voice quality, so once again there is a tradeoff between security and voice quality, and a need for speed. Fortunately, the difficulties are not insurmountable. NIST-sponsored testing [29] has shown that IPsec can be incorporated into a SIP network with roughly a three-second additional delay in call setup times, an acceptable delay for many applications. This section explains the issues involved in successfully incorporating IPsec encryption into VOIP services.

8.1 IPsec

IPsec is the preferred form of VPN tunneling across the Internet. There are two basic protocols defined in IPsec: Encapsulating Security Payload (ESP) and Authentication Header (AH) (see Figure 11). Both schemes provide connectionless integrity, source authentication, and an anti-replay service [30]. The tradeoff between ESP and AH is the increased latency in the encryption and decryption of data in ESP and a "narrower" authentication in ESP, which normally does not protect the IP header "outside" the ESP header [30], although IKE can be used to negotiate the security association (SA), which includes the secret symmetric keys. In this case, the addresses in the header (transport mode) or new/outer header (tunnel mode) are indirectly protected, since only the entity that negotiated the SA can encrypt/decrypt or authenticate the packets.. Both schemes insert an IPsec header (and optionally other data) into the packet for purposes, such as authentication.

IPsec also supports two modes of delivery: Transport and Tunnel. Transport mode encrypts the payload (data) and upper layer headers in the IP packet. The

IP header and the new IPsec header are left in plain sight. So if an attacker were to intercept an IPsec packet in transport mode, they could not determine what it contained; but they could tell where it was headed, allowing rudimentary traffic analysis. On a network entirely devoted to VOIP, this would equate to logging which parties were calling each other, when, and for how long. Tunnel mode encrypts the entire IP datagram and places it in a new IP Packet. Both the payload and the IP header are encrypted. The IPsec header and the new IP Header for this encapsulating packet are the only information left in the clear. Usually each "tunnel" is between two network elements such as a router or a gateway. In some cases, such as for mobile users, the tunnel could be between a router/gateway on one end and a client on the other end. The IP addresses of these nodes are used as the unencrypted IP address at each hop. Hence, at no point is a plain IP header sent out containing both the source and destination IP. Thus if an attacker were to intercept such packets, they would be unable to discern the packet contents or the origin and destination. Note that some traffic analysis is possible even in tunnel mode, because gateway addresses are readable. If a gateway is used exclusively by a particular organization, an attacker can determine the identity of one or both communicating organizations from the gateway addresses. IPsec allows nodes in the network to negotiate not only a security policy, which defines the security protocol and transport mode as described previously, but also a security association defining the encryption algorithm and algorithm key to be used

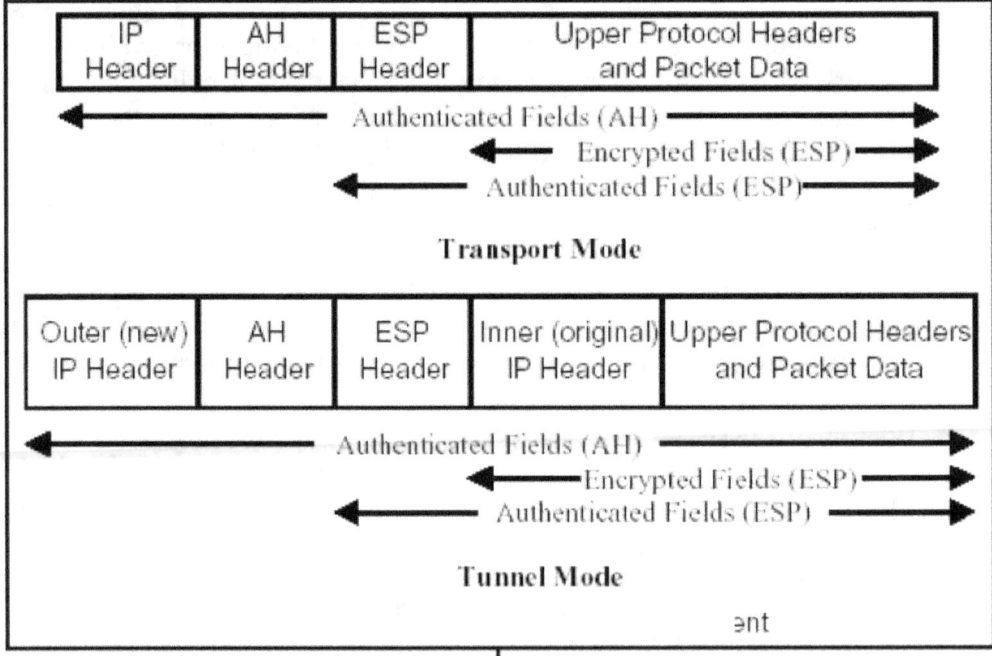

Figure 11. IPsec Tunnel and Transport Modes

8.2 The Role of IPsec in VOIP

The prevalence and ease of packet sniffing and other techniques for capturing packets on an IP based network makes encryption a necessity for VOIP. Security in VOIP is concerned both with protecting what a person says as well as to whom the person is speaking. IPsec can be used to achieve both of these goals as long as it is applied with ESP using the tunnel method. This secures the identities of both the endpoints and protects the voice data from prohibited users once packets leave the corporate intranet. The incorporation of IPsec into IPv6 will increase the availability of encryption, although there are other ways to secure this data at the application level. VOIPsec (VOIP using IPsec) helps reduce the threat of man in the middle attacks, packet sniffers, and many types of voice traffic analysis. Combined with the firewall implementations in the previous chapter, IPsec makes VOIP more secure than a standard phone line, where people generally assume the need for physical access to tap a phone line is deterrent enough. It is important to note, however, that IPsec is not always a good fit for some applications, so some protocols will continue to rely on their own security features.

8.3 Local VPN Tunnels

Virtual Private Networks (VPNs) are "tunnels" between two endpoints that allow for data to be securely transmitted between the nodes. The IPsec ESP tunnel is a specific kind of VPN used to traverse a public domain (the Internet) in a private manner. Many implementations of VOIP have attempted to make use of other VPN techniques, including VPN tunneling within an organization's's intranet. The use and benefits of VPNs in IPsec have been great enough for some to claim "VOIP is the killer app for VPNs" [26]. VPN tunnels within a corporate LAN or WAN are much more secure and generally faster than the IPsec VPNs across the Internet because data never traverses the public domain, but they are not scaleable. This sort of implementation has a physical limit at the size of the private network, and as VOIP becomes more widely spread, it is not practical for an implementation to regard calls outside the local network as a black box. Also, no matter how the VPN is set up, the same types of attacks and issues associated with IPsec VPNs are applicable, so we consider here only the case of IPsec tunneling and assume the security solutions can be scaled down to an internal network if needed.

8.4 Difficulties Arising from VOIPsec

IPsec has been included in IPv6. It is a reliable, robust, and widely implemented method of protecting data and authenticating the sender. However, there are several issues associated with VOIP that are not applicable to normal data traffic. Of particular interest are the Quality of Service (QoS) issues discussed in Chapter 3. Chief among these are latency, jitter, and packet loss. These issues are introduced into the VOIP environment because it is a real time media transfer, with only 150 ms to deliver each packet. In standard data transfer over TCP, if a packet is lost, it can be resent by request. In VOIP, there is no time to do this.

Packets must arrive at their destination and they must arrive fast. Of course the packets must also be secure during their travels, thus the introduction of VOIPsec. However, the price of this security is a decisive drop in QoS caused by a number of factors.

A 2002 study by researchers at the University of Milan [8] focused on the effect of VOIPsec on various QoS issues and on the use of header compression as a solution to these problems. They studied several codecs, encryption algorithms, and traffic patterns to garner a broad description of these effects. Their results are an integral part of the next few sections.

Some empirical results developed by Cisco are available as well [31]
- Delay
    - Processing—PCM to G.729 to packet
    - Encryption — ESP encapsulation + 3DES
    - Serialization — time it takes to get a packet out of the router, each "hop" generally has fixed delay.
        - IPsec overhead: about 40 bytes (depending configuration)
        - IP header: 20 bytes
        - UDP + RTP headers: 20 bytes
        - RTP header compression: 3 bytes for IP+UDP+RTP
- Effects on 8 kbps CODEC (voice data: 20 bytes)
    - clear text voice has an overhead of 3 bytes, which suggests required bandwidth of approximately 9 kbps
    - IPsec encrypted voice: overhead 80 bytes and required bandwidth 40 kbps

8.5 Encryption / Decryption Latency

The studies performed by Barbieri et al. revealed the cryptographic engine as a bottleneck for voice traffic transmitted over IPsec. The driving factor in the degraded performance produced by the cryptography was the scheduling algorithms in the crypto-engine itself, which will be covered in Section 8.6. However, there still was significant latency due to the actual encryption and decryption. Barbieri et al. set up a controlled experiment to measure the effect of encryption and decryption on throughput. They tested four cryptographic algorithms on a fully VOIP dedicated network with a 100Mbps link (to negate saturation issues) using the same traffic in plain form as a benchmark. The algorithms tested were (in increasing order of computational expense) DES, 3DES, NULL (no encryption) + SHA-1, and 3DES + SHA-1. The results showed that the computationally lighter algorithms achieved better throughput than the more expensive ones. The disparities between each of the algorithms represent the relative latencies associated with the computational time for each algorithm. The range in throughput is significant, with a difference of approximately 500 packets per second between DES and 3DES + SHA-1 at a high traffic volume.

Encryption/decryption latency is a problem for any cryptographic protocol, because much of it results from the computation time required by the underlying encryption. With VOIP's use of small packets at a fast rate and intolerance for packet loss, maximizing throughput is critical. However
, this comes with a price, because although DES is the fastest of these encryption algorithms, it is also the easiest to crack. Current rules prohibit the use of DES for protection of US Government information. Thus, designers are once again forced to toe the line between security and voice quality. Two solutions to this problem are using faster encryption algorithms (9.3) and incorporating QoS into the crypto-engine (9.4). Latency is less of a problem for management and/or signaling data than for voice channel traffic.

8.6 Scheduling and the Lack of QoS in the Crypto-Engine

The crypto-engine is a severe bottleneck in the VOIP network. As just noted, the encryption process has a debilitating effect on QoS, but this is not the highest degree factor in the slowdown. Instead, the driving force behind the latency associated with the crypto-engine is the scheduling algorithm for packets that entered the encryption/decryption process. While routers and firewalls take advantage of QoS to determine priorities for packets, crypto-engines provide no support for manual manipulation of the scheduling criteria. In ordinary data traffic this is less of an issue because inordinately more packets pass through the router than the crypto-engine, and time is not as essential. But in VOIP, a voluminous number of small packets must pass through both the crypto engine and the router. Considering the time urgency issues of VOIP, the standard FIFO scheduling algorithm employed in today's crypto-engines creates a severe QoS issue.

Barbieri et al. found that the throughput of the crypto scheduling algorithm actually increased with larger packet sizes. They concluded that scheduling a greater number of packets had a more degrading effect on performance than encrypting/decrypting fewer (but larger) packets. So with the numerous small packets VOIP uses, the crypto engine soon reaches the saturation point, and throughput is compromised. This accounts for the asymptotic behavior of the throughput for encrypted traffic against plain traffic's continuous rise and the increased delay of encrypted traffic that Barbieri et al. observed.

These QoS violations derived from the crypto-engine are exacerbated by the presence of actual data traffic on a VOIP network. Since one of the primary motivations for the development of VOIP is the ability of voice and data to share the same network, this scenario is to be expected. Barbieri et al.'s experiments showed that such a combination of traffic has disastrous effects on VOIPsec. This is especially true if the heterogeneous data needs to be encrypted and decrypted. Since the crypto-engine has no functionality for changing its own priority schema based on the type of traffic it is presented with, the VOIP packets are at the mercy of the FIFO scheduling algorithm, and are often left waiting behind larger heterogeneous packets, despite the lesser urgency of these large packets. The non-uniform pattern of data traffic also contributes to a great deal of jitter in VOIP. The variation in the bandwidth usage caused by heterogeneous packets wreaks havoc with the delay times of the fairly uniform VOIP packets, causing them to arrive in spurts. If these spurts exceed the timing mechanism associated with the buffer, then packet loss can occur. The development of VOIP-aware crypto schedulers would help to relieve this problem.

8.7 Expanded Packet Size

IPsec also increases the size of packets in VOIP, which leads to more QoS issues. It has been shown that increased packet size increases throughput through the crypto-engine, but to conclude from this that increased packet size due to IPsec leads to better throughput would be fallacious. The difference is that the increase in packet size due to IPsec does not result in an increased payload capacity. The increase is actually just an increase in the header size due to the encryption and encapsulation of the old IP header and the introduction of the new IP header and encryption information. (see 8.1). This leads to several complications when IPsec is applied to VOIP. First, the effective bandwidth is decreased as much as 63% [9]. Thus connections to single users in low bandwidth areas (i.e. via modem) may become infeasible. The bandwidth performance reductions for various encryption algorithms are presented in [8]. The size discrepancy can also cause latency and jitter issues as packets are delayed by decreased network throughput or bottlenecked at hub nodes on the network (such as routers or firewalls).

8.8 IPsec and NAT Incompatibility

IPsec and NAT compatibility is far from ideal. NAT traversal completely invalidates the purpose of AH because the source address of the machine behind the NAT is masked from the outside world. Thus, there is no way to authenticate the true sender of the data. The same reasoning demonstrates the inoperability of source authentication in ESP. We have defined this as an essential feature of VOIPsec, so this is a serious problem. There are several other issues that arise when ESP traffic attempts to cross a NAT. If only one of the endpoints is behind a NAT, the situation is easier [32] If both are behind NATs, IKE negotiation can be used for NAT traversal, with UDP encapsulation of the IPsec packets.

9 Solutions to the VOIPsec Issues

Thus far, we have raised a number of significant concerns with IPsec's role in VOIP. However, many of these technical problems are solvable. Despite the difficulty associated with these solutions it is well worth the establishment of a secure implementation of VOIPsec.

9.1 Encryption at the End Points

One proposed solution to the bottlenecking at the routers due to the encryption issues is to handle encryption/decryption solely at the endpoints in the VOIP network [33]. One consideration with this method is that the endpoints must be computationally powerful enough to handle the encryption mechanism. But typically endpoints are less powerful than gateways, which can leverage hardware acceleration across multiple clients. Though ideally encryption should be maintained at every hop in a VOIP packet's lifetime, this may not be feasible with simple IP phones with little in the way of software or computational power. In such cases, it may be preferable for the data be encrypted between the endpoint and the router (or vice versa) but unencrypted traffic on the LAN is slightly less damaging than unencrypted traffic across the Internet. Fortunately, the increased processing power of newer phones is making endpoint encryption less of an issue. In addition, SRTP and MIKEY are future protocols for media encryption and key management enabling secure interworking between H.323 and SIP based clients.

9.2 Secure Real Time Protocol (SRTP)

RTP (Real-time Transport Protocol) is commonly used for the transmission of real-time audio/video data in Internet telephony applications. Without protection RTP is considered insecure, as a telephone conversation over IP can easily be eavesdropped. Additionally, manipulation and replay of RTP data could lead to poor voice quality due to jamming of the audio/video stream. Modified RTCP (Real-time Transport Control Protocol) data could even lead to an unauthorized change of negotiated quality of service and disrupt the processing of the RTP stream.

The Secure Real-time Protocol is a profile of the Real-time Transport Protocol (RTP) offering not only confidentiality, but also message authentication, and replay protection for the RTP traffic as well as RTCP (Real-time Transport Control Protocol). SRTP was being standardized at the IETF in the AVT working group. It was released as RFC 3711 in March 2004.

SRTP provides a framework for encryption and message authentication of RTP and RTCP streams. SRTP can achieve high throughput and low packet expansion. SRTP is independent of a specific RTP stack implementation and of a specific key management standard, but Multimedia Internet Keying (MIKEY) has been designed to work with SRTP.

AES in counter mode is the default algorithm, if encryption is desired. AES-f8 mode is an option for UMTS applications. The pre-defined authentication transform is HMAC-SHA1. The default session authentication key-length is 160 bits, the default authentication tag length is 80 bits. The key derivation function is AES in counter mode with a 128-bit master key from the key management. Interface for hardware-crypto support (e.g. IP phones). In comparison to the security options for RTP there are some advantages to using SRTP. The advantages over the RTP standard security and also over the H.235 security for media stream data are listed below.

SRTP provides increased security, achieved by
- Confidentiality for RTP as well as for RTCP by encryption of the respective payloads;
- Integrity for the entire RTP and RTCP packets, together with replay protection;
- The possibility to refresh the session keys periodically, which limits the amount of cipher text produced by a fixed key, available for an adversary to cryptanalyze;
- An extensible framework that permits upgrading with new cryptographic algorithms;
- A secure session key derivation with a pseudo-random function at both ends;
- The usage of salting keys to protect against pre-computation attacks;
- Security for unicast and multicast RTP applications.

SRTP has improved performance attained by
- Low computational cost asserted by pre-defined algorithms;
- Low bandwidth cost and a high throughput by limited packet expansion and by a framework preserving RTP header compression efficiency;
- Small footprint that is a small code size and data memory for keying information and replay lists.

The following characteristics also argue for SRTP:
- It is defined as a profile of RTP, so that it can be easily integrated into existing RTP stacks. For example SRTP may use RTP padding because the encrypted portion is the exact size of the plaintext for the pre-defined algorithms.
- It provides independence from the underlying transport, network, and physical layers used by RTP, in particular high tolerance to packet loss and re-ordering, and robustness to transmission bit-errors in the encrypted payload.
- It lightens the burden of the key management due to the fact that a single master key can provide keying material for confidentiality and integrity protection, both for the SRTP stream and the corresponding SRTCP stream. For special requirements a single master key can protect several

SRTP streams.

Because SRTP is defined as an RTP profile it may be used with existing multimedia standards. H.323 SRTP support is defined within the H.235 Annex G (currently in draft status), for SIP or more precisely SDP enhancements have been defined to transport the key management data necessary for SRTP. Thus, the combination of SRTP and MIKEY may be used to provide end-to-end encryption even between different multimedia signaling standards like H.323 and SIP.

9.3 Key Management for SRTP – MIKEY

SRTP uses a set of negotiated parameters from which session keys for encryption, authentication and integrity protection are derived. MIKEY (Section 4.2.2.2) describes a key management scheme that addresses real-time multimedia scenarios (e.g. SIP calls and RTSP sessions, streaming, unicast, groups, multicast) and is currently being standardized within the IETF's MSEC group. The focus lies on the setup of a security association for secure multimedia sessions including key management and update, security policy data, etc., such that requirements in a heterogeneous environment are fulfilled. MIKEY also supports the negotiation of single and multiple crypto sessions. This is especially useful for the case where the key management is applied to SRTP, since here RTP and RTCP may to be secured independently. Deployment scenarios for MIKEY comprise peer-to-peer, simple one-to-many, and small-size interactive group scenarios.

MIKEY supports the negotiation of cryptographic keys and security parameters (SP) for one or more security protocols. This results in the concept of crypto session bundles, which describe a collection of crypto sessions that may have a common Traffic Encryption Key (TEK) Generation Key (TGK) and belonging session security parameters.

MIKEY has some important properties:
o MIKEY can be implemented as an independent software library to be easily integrated in a multimedia communication protocol. It offers independency of a specific communication protocol (SIP, H.323, etc.)
o Establishment of key material within a 2-way handshake, therefore best suited for real-time multimedia scenarios
o There are four options for Key Distribution:
    o Preshared-key
    o Public-key encryption
    o Diffie-Hellman key exchange protected by public-key encryption
    o Diffie-Hellman key exchange protected with preshared-key and keyed hash functions (using an MIKEY extension (DHHMAC))
o Re-keying Support
o Multicast Support (one sender)

9.4 Better Scheduling Schemes

The incorporation of AES or some other speedy encryption algorithm could help temporarily alleviate the bottleneck, but this is not a scalable solution because it does not address the highest degree cause of the slowdown. Without a way for the crypto-engine to prioritize packets, the engine will still be susceptible to DoS attacks and starvation from data traffic impeding the time-urgent VOIP traffic. A few large packets can clog the queue long enough to make the VOIP packets over 150 ms late (sometimes called head-of-line blocking), effectively destroying the call. Ideally, the crypto-engine would implement QoS scheduling to favor the voice packets, but this is not a realistic scenario due to speed and compactness constraints on the crypto-engine. One solution implemented in the latest routers is to schedule the packets with QoS in mind prior to the encryption phase. Although this heuristic solves the problem for all packet poised to enter the crypto engine at a given time, it does not address the problem of VOIP packets arriving at a crypto–engine queue that is already saturated with previously scheduled data packets. QoS prioritizing can also be done after the encryption process provided your encryption procedures preserve the ToS bits from the original IP header in the new IPsec header. This functionality is not guaranteed and is dependent on one's network hardware and software, but if it is implemented it allows for QoS scheduling to be used at every hop the encrypted packets encounter. There are security concerns any time information on the contents of a packet is left in the clear, including this ToS-forwarding scheme, but with the sending and receiving addresses concealed, this is not as egregious as a cursory glance would make it seem. Still neither the pre-encryption or post-encryption schemes actually implement QoS or any other prioritizing scheme to enhance the crypto-engine's FIFO scheduler. Speed and compactness constraints on this device may not allow such algorithms to be applied for some time.

9.5 Compression of Packet Size

A novel approach to the QoS issues associated with VOIPsec is proposed by Barbieri et al. at the conclusion of their studies of VOIPsec traffic. Their solution targets the increase in packet size stemming from the use of IPsec. They implemented cIPsec: a version of IPsec that compresses the internal header of a packet down to approximately four bytes. This is possible because much of the data in the internal headers of a packet remained constant or was duplicated in the outer header.

The initial test results reported from the University of Milan indicate that the compression of IPsec headers results in bandwidth usage comparable to that of plain IP. This in turn results in considerably less jitter, latency, and better crypto-engine performance. The crypto-engine performance also improves. There is, of course, a price for these speedups. The compression scheme puts more strain on the CPU and memory capabilities of the endpoints in order to achieve the compression, and, of course, both ends of a connection must use the same compression algorithm. However, the study found that the time lost to compression was made up for at the encryption phase, as the crypto-engine is more efficient with the compressed packets. One thing they did not consider is the tremendous strain put on end-point CPU's as opposed to the crypto-engine. The endpoint CPU may be computationally slow (in the case of a simple VOIP phone) or may be performing many more operations than just VOIP (in the case of a PC-based phone). In either case, the actual time required to perform the compression may take much longer than the time saved in the crypto-engine. It remains to be seen if this is the case, as Barbieri's model of cIPsec was not tested under high CPU load conditions.

It is important to note that the compression scheme used in cIPsec only compresses the packet header information. The compression QoS issues associated with audio codecs are not applicable in this scenario because no actual media is being condensed, only the IP headers. However, packet loss does have an exacerbated detrimental effect (for a different reason) on packets compressed under the cIPsec scheme. Barbieri's scheme needs to maintain information at the system endpoints regarding the current session. When packets are lost, they cannot be re-sent and the endpoints need to resynchronize [9]. However, the time saved in the crypto-engine and the security provided may be well worth this price of this approach. Further testing will be required to determine the validity of this solution under diverse network and environmental conditions.

9.6 Resolving NAT/IPsec Incompatibilities

There are solutions to the IPsec/NAT incompatibility problem previously outlined in section 8.2. Straka [32] discusses several of these, including Realm-Specific IP RSIP), IPv6 Tunnel Broker, IP Next Layer (IPNL), and UDP encapsulation. RSIP is designed as a replacement for NAT and provides a clear tunnel between hosts and the RSIP Gateway. RSIP supports both AH and ESP, but implementing RSIP would require a significant overhaul of the current LAN architecture so while it is quite an elegant solution, it is currently infeasible. Perhaps as a result of these problems, RSIP is not widely used. The IPv6 tunnel broker method uses an IPv6 tunnel as an IPsec tunnel, and encapsulates an IPv6 packet in an IPv4 packet. But this solution also requires LAN upgrades and doesn't work in situations where multiple NATs are used. IPNL introduces a new layer into the network protocols between IP and TCP/UDP to solve the problem, but IPNL is in competition with IPv6 and IPv6 is a much more widely used standard.

The most likely widespread solution to the problem of NAT traversal is UDP encapsulation of IPsec. This implementation is supported by the IETF and effectively allows all ESP traffic to traverse the NAT. In tunnel mode, this model wraps the encrypted IPsec packet in a UDP packet with a new IP header and a new UDP header, usually using port 500. This port was chosen because it is currently used by IKE peers to communicate so overloading the port does not require any new holes to be punched in the firewall [34]. The SPI field within the

UDP-encapsulated packet is set to zero to differentiate it from an actual IKE communication. This solution allows IPsec packets to traverse standard NATs in both directions. The adoption of this standard method should allow VOIPsec traffic to traverse NATs cleanly, although some extra overhead is added in the encapsulation/decapsulation process. IKE negotiation will also be required to allow for NAT traversal. The problem still remains that IP-based authentication of the packets cannot be assured across the NAT, (although fully qualified domain names could be used) but the use of a shared secret (symmetric key) negotiated through IKE could provide authentication. It is important to note that IP-based authentication is weak compared with methods using cryptographic protocols.

10 Planning for VOIP Deployment

VOIP is still an emerging technology, so it is somewhat speculative to develop a complete picture of what a mature worldwide VOIP network will one day look like. As the emergence of SIP has shown, new technologies and new protocol designs have the ability to change VOIP. The situation is analogous to the state of the Internet in the late 80s and early 90s. Competing protocols and designs for the infrastructure of the net flourished at the time, but as the purpose of the Internet became more defined with the emergence of the world wide web and other staples of today's net, the structure and protocols became standardized and interoperability became much easier. The same may one day be true of VOIP. Although there are currently many different architectures and protocols to choose from, eventually a dominant standard will emerge.

The most obvious of these competing standards are SIP and H.323. Comparisons are made in numerous academic papers including [16, 18, and 15]. SIP is a fast growing protocol with similarities to current Internet standards such as HTTP, but it has yet to reach the level of deployment of H.323 [15]. The opinion of many academics [16, 18] seems to favor SIP, and we have seen that some of the security issues associated with VOIP become simpler with the SIP scheme. It is not clear which standard will prevail in the market. It is misleading to portray the choice between SIP and H.323 as mutually exclusive. In fact, in today's non-standardized VOIP environment, organizations looking to integrate several VOIP networks ought to support both protocols. Several companies have developed infrastructure elements to enable multi-protocol telephony. As voice and data networks converge, support for both protocols is essential for a robust and forward-looking network. Although the future will probably see the emergence of one of these protocols as the defined standard in the field the present disorganization makes support for both protocols in a VOIP network a pertinent issue. Deploying a VOIP network in today's non-standardized world requires support for both protocols. Consequently, organizations moving to VOIP should seek out gateways and other network elements that can support both H.323 and SIP. Such a strategy helps to ensure a stable and robust VOIP network in the years that come, no matter which protocol prevails.

The other high-level issue in VOIP security today is the choice of end-to-end VPNs versus firewall-based VPNs. That is, VOIP traffic must traverse firewalls one way or the other. The question becomes, should one build firewalls with ALGs, proxies, firewall control proxies, and IPsec functionality to facilitate this, or simply tunnel all VOIP traffic straight through the firewall with a VPN. The security benefits and administrative troubles associated with each of these implementations have been presented in detail in this publication. The use of VPNs has been touted by many industry articles as the definitive solution to the tribulations posed by firewall and NAT traversal in tunnel mode. However, much of their research has focused on small-scale operations where VOIP phones are not used in the volume needed to overwhelm the crypto-engines or congest the network enough to cause a significant downturn in QoS. Thus it is not clear that the large-scale implementation of a VPN tunneling system for VOIP is an effective solution for an enterprise network [see Broadcom in 27], as there is significant overhead associated with VOIPsec. Also, the ability to use firewalls for analyzing VOIP traffic for malicious or suspicious patterns would be lost. This being said, the implementation of VOIP aware firewalls and proxies incurs a significant cost now and in the future. Such protocol specific hardware would need to be upgraded each time standards evolve. A third solution that has not been fully developed yet is a hybrid system, where call setup information (H.323 or SIP) is sent through a VOIP aware gateway/firewall but the RTP traffic itself is encrypted and tunneled over VPN. The call setup protocols could be secured using their proprietary authentication mechanisms [17] in place of the IPsec tunnel. This would seemingly combine the network protection of the firewall and the data security/protection of IPsec. Also, the robust authentication mechanisms and abstraction of the voice network from the data network accomplished by SIP proxies and H.323 gateways/gatekeepers would be preserved. However, no expanded study has been done on the ramifications of this hybrid approach.

There are a number of privacy issues regarding storage of call detail records. Agencies and other organizations should review these issues with their legal advisors. US Government guidance on these issues includes the following:

- the Privacy Act of 1974.

- Office of Management and Budget "*Guidance on the Privacy Act Implications of Call Detail Programs to Manage Employees' Use of the Government's Telecommunication System*" (See FEDERAL REGISTER, 52 FR 12990, April 20, 1987).

- NARA General Records Schedule 12, which requires a 36-month retention of telephone CDR records
  http://www.archives.gov/records_management/ardor/grs12.html

- 21 CFR 102-172, Federal Management Regulation (FMR), Telecommunications Management Policy. 21 CFR 102-172 . Note that 21 CFR 102-172 replaces 21 CFR 101-35, Federal Property Management Regulation (FPMR), Telecommunications Management Policy, which expired in August 2001.

The construction of a VOIP network is an intricate procedure that should be studied in great detail before being attempted. New risks can be introduced, and vulnerabilities of data packet networks appear in new guises for VOIP (see Appendix A for more detailed discussion of vulnerabilities of VOIP and their relation to data network vulnerabilities). The integration of a VOIP system into an already congested or overburdened network could be catastrophic for an organization's technology infrastructure. There is no easy "one size fits all" solution to the issues discussed in these chapters. The use of VPNs, vs. ALG-like solutions and the choice of SIP or H.323 are decisions that must be made based on the specific nature of the current network and the VOIP network to be installed.

VOIP can be done securely, but the path is not smooth. It will likely be several years before standards issues are settled and VOIP systems become a mainstream commodity. Until then, organizations must proceed cautiously, and not assume that VOIP components are just more peripherals for the local network. Above all, it is important to keep in mind the unique requirements of VOIP, acquiring the right hardware and software to meet the challenges of VOIP security.

REFERENCES

1 F. Robles. "The VOIP Dilemma", SANS Institute, http://www.sans.org/rr/whitepapers/voip/1452.php

2. National Institute of Standards and Technology, DRAFT FIPS Publication 199, *Standards for Security Categorization of Federal Information and Information Systems,* September 18, 2003   http://csrc.nist.gov/publications/drafts/draft-fips-pub-199.pdf

3. W.C. Hardy, QoS Measurement and Evaluation of Teleocmmunication Quality of Service, John Wiley & Sons, 2001.

4. W.C. Hardy, VOIP Service Quality: Measuring and Evaluating Packet-Switched Voice, McGraw-Hill, 2003.

5. International Telecommunications Union. ITU-T Recommendation G.114 (1998): "Delay".

6. P. Mehta and S. Udani, "Overview of Voice over IP". Technical Report MS-CIS-01-31, Department of Computer Information Science, University of Pennsylvania, February 2001.

7. B. Goode, "Voice Over Internet Protocol (VOIP)". *Proceedings of thee IEEE*, VOL. 90, NO. 9, Sept. 2002.

8. R. Barbieri, D. Bruschi, E Rosti, "Voice over IPsec: Analysis and Solutions". *Proceedings of the 18th Annual Computer Security Applications Conference*,2002.

9. C-N. Chuah, "Providing End-to-End QoS for IP based Latency sensitive Applications.". Technical Report, Dept. of Electrical Engineering and Computer Science, University of California at Berkeley, 2000.

10. B. Goode, "Voice Over Internet Protocol (VOIP)". *Proceedings of thee IEEE*, VOL. 90, NO. 9, Sept. 2002.

11. Anonymous, "Voice Over IP Via Virtual Private Networks: An Overview". White Paper, AVAYA Communication, Feb. 2001.

12. R. Sinden, "Comparison of Voice over IP with circuit switching techniques". Department of electronics and Computer Science, Southampton University, UK, Jan. 2002.

13. K. Percy and M. Hommer, "Tips from the trenches on VOIP". *Network World Fusion,* Jan. 2003

14. J. Larson, T. Dawson, M. Evans, J.C. Straley, "Defending VoIP Networks from DDoS Attacks", *GlobeCom 2004 VoIP Security Workshop.*

15. Anonymous, "H.323 and SIP Integration". White Paper, Cisco Systems, 2001.

16. K. Siddiqui, M. Kamran, S. Tajammul, "Comparison of H.323 and SIP for IP Telephony Signaling". In *Proceedings of IEEE 4th International Multioptics Conference,* Lahore, Pakistan, Dec. 2001.

17. J. Thalhammer, "Security in VOIP-Telephony Systems". Master Thesis, Institute for Applied Information Processing and Communications, Graz U. of Technology,

18. H. Schulzrinne and J. Rosenberg. A comparison of SIP and H.323 for Internet telephony. In *Proc. International Workshopon Network and Operating System Support for Digital Audio and Video (NOSSDAV)*, Cambridge, England, July1998.

19. M. Marjalaakso, "Security Requirements and Constraints of VOIP". Department of Eleectrical Engineering and Telecommunications, Helsinki University of Technology, 2001.

20 Jablon, "Strong Password-Only Authenticated **Key Exchange**", Mar. 1997, http://www.integritysciences.com/speke97.html

21. Anonymous, "H.323 and firewalls: The problems and Pitfalls of Getting H.323 safely through firewalls" Developer note, Intel Corporation, Apr. 1997.

22. J. Rosenberg and H. Schulzrinne, "SIP Traversal through Residential and Enterprise NATs and Firewalls". Internet Draft, Internet Engineering Task Force, Mar. 2001.

23. J. Halpern, "IP Telephony Security in Depth". White Paper, Cysco Systems, 2002.

24. Anonymous, "Voice Over IP Via Virtual Private Networks: An Overview". White Paper, AVAYA Communication, Feb. 2001.

25. A. Romeo, G. Romolotti, M. Mattavelli, D. Mlynek, "Cryptosystem Architectures for very High Throughput Multimedia Encryption: The RPK Solution", Swiss Federal Institute of technology, Lausanne, Switzerland, 1999.

26. A. Conry-Murray, "Emerging Technology: Security and Voice over IP – Let's Talk". Commweb, Nov. 2002.

27. P. Hochmuth and T. Greene, "Firewall limits vex VOIP users". *Network World Fusion*, July 2002.

28. Anonymous, "Traversing Firewalls and NATs With Voice and Video Over IP: An Examination of the Firewall/NAT Problem, Traversal Methods, and their Pros and Cons" . Wainhouse Research, Apr. 2002.

29. Telcordia Technologies, "Performance and Security Analysis of SIP using IPsec", National Institute of Standards and Technology, January, 2004.

30. G. Egeland, "Introduction to IPsec in IPv6". Eurescom, http://www.eurescom.de/~publicwebdeliverables/P1100series/P1113/D1/pdfs/pir1/41_IPsec_intro.pdf

31. Cisco Networkers 2000, http://www.cisco.com/networkers/nw00/pres/2403.pdf

32. R. Straka, "IPsec VPN Traffic and NAT / NAPT Traversal: Searching for a Universal Solution". Center for Information Security, University of Tulsa, Dec. 2002.

33. O. Arkin, "Why E.T. Can't Phone Home?: Security Risk Factors with IP Telephony based Networks" . Sys-Security Group, Nov. 2002.

34. L. Phifer, "Slipping IPsec Past NAT". ISP Planet, Apr. 2001.

34. Internet Draft: End-to-middle Security in the Session Initiation Protocol (SIP), K.Ono, S. Tachimoto, February 2004, Work in Progress http://www.ietf.org/Internet-drafts/draft-ono-sipping-end2middle-security-01.txt

35. Internet Draft: A Mechanism to Secure SIP information inserted by Intermediaries,M.Barnes, October 2003, Work in Progress http://search.ietf.org/Internet-drafts/draft-barnes-sipping-sec-inserted-info-01.txt

36. Request for Comments 3489: STUN: Simple Traversal of UDP through Network Address Translators, , J.Rosenberg, et al, March 2003. http://www.ietf.org/rfc/rfc3489.txt

A Appendix: VOIP Risks, Threats, and Vulnerabilities

This appendix details some of the potential threats and vulnerabilities in a VOIP environment, including vulnerabilities of both VOIP phones and switches. Threat discussion is included because the varieties of threats faced by an organization determine the priorities in securing its communications equipment. Not all threats are present in all organizations. A commercial firm may be concerned primarily with toll fraud, while a government agency may need to prevent disclosure of sensitive information because of privacy or national security concerns. Information security risks can be broadly categorized into the following three types: *confidentiality, integrity,* and *availability*, (which can be remembered with the mnemonic "CIA"). Additional risks relevant to switches are fraud and risk of physical damage to the switch, physical network, or telephone extensions.

Packet networks depend for their successful operation on a large number of configurable parameters: IP and MAC (physical) addresses of voice terminals, addresses of routers and firewalls, and VOIP specific software such as Call Managers and other programs used to place and route calls. Many of these network parameters are established dynamically every time a network component is restarted, or when a VOIP telephone is restarted or added to the network. Because there are so many places in a network with dynamically configurable parameters, intruders have a wide array of potentially vulnerable points to attack.

Vulnerabilities described in this section are generic and may not apply to all systems, but investigations by NIST and other organizations have found these vulnerabilities in a number of VOIP systems. In addition, this list is not exhaustive; systems may have security weaknesses that are not included in the list. For each potential vulnerability, a recommendation is included to eliminate or reduce the risk of compromise.

A.1 Confidentiality and Privacy

Confidentiality refers to the need to keep information secure and private. For home computer users, this category includes confidential memoranda, financial information, and security information such as passwords. In a telecommunications switch, eavesdropping on conversations is an obvious concern, but the confidentiality of other information on the switch must be protected to defend against toll fraud, voice and data interception, and denial of service attacks. Network IP addresses, operating system type, telephone extension to IP address mappings, and communication protocols are all examples of information that, while not critical as individual pieces of data, can make an attacker's job easier

With conventional telephones, eavesdropping usually requires either physical access to tap a line, or penetration of a switch. Attempting physical access increases the intruder's risk of being discovered, and conventional PBXs have fewer points of access than VOIP systems. With VOIP, opportunities for eavesdroppers increase dramatically, because of the many nodes in a packet network.

Switch Default Password Vulnerability

It is common for switches to have a default login/password set, e.g., admin/admin, or root /root. This vulnerability also allows for wiretapping conversations on the network with port mirroring or bridging. An attacker with access to the switch administrative interface can mirror all packets on one port to another, allowing the indirect and unnoticeable interception of all communications. Failing to change default passwords is one of the most common errors made by inexperienced users. If possible, remote access to the graphical user interface should be disabled to prevent the interception of plaintext administration sessions. Some devices provide the option of a direct USB connection in addition to remote access through a web browser interface. Disabling port mirroring on the switch should also be considered.

Classical Wiretap Vulnerability

Attaching a packet capture tool or protocol analyzer to the VOIP network segment makes it easy to intercept voice traffic.

A good physical security policy for the deployment environment is a general first step to maintaining confidentiality. Disabling the hubs on IP Phones as well as developing an alarm system for notifying the administrator when an IP Phone has been disconnected will allow for the possible detection of this kind of attack.

ARP Cache Poisoning and ARP Floods

Because many systems have little authentication, an intruder may be able to log onto a computer on the VOIP network segment, and then send ARP commands corrupting ARP caches on sender(s) of desired traffic, then activate IP. An ARP flood attack on the switch could render the network vulnerable to conversation eavesdropping. Broadcasting ARP replies blind is sufficient to corrupt many ARP caches.

Corrupting the ARP cache makes it possible to re-route traffic to intercept voice and data traffic. Use authentication mechanisms provided wherever possible and limit physical access to the VOIP network segment.

Web Server interfaces

Both VOIP switches and voice terminals are likely to have a web server interface for remote or local administration. An attacker may be able to sniff plaintext HTTP packets to gain confidential information. This would require access to the local network on which the server resides. If possible, do not use an HTTP server. If it is necessary to use a web server for remote administration, use the more secure HTTPS (HTTP over SSL or TLS) protocol.

IP Phone Netmask Vulnerability

A similar effect of the ARP Cache Vulnerability can be achieved by assigning a subnet mask and router address to the phone crafted to cause most or all of the packets it transmits to be sent to an attacker's MAC address. Again, standard (1q aware) IP forwarding makes the intrusion all but undetectable.

A firewall filtering mechanism can reduce the probability of this attack. Remote access to IP phones is a severe risk.

Extension to IP Address Mapping Vulnerability

Discovering the IP address corresponding to any extension requires only calling that extension and getting an answer. A protocol analyzer or packet capture tool attached to the hub on the dialing instrument will see packets directly from the target instrument once the call is answered. Knowing the IP address of a particular extension is not a compromise in itself, but makes it easier to accomplish other attacks. For example, if the attacker is able to sniff packets on the local network used by the switch, it will be easy to pick out packets sent and received by a target phone. Without knowledge of the IP address of the target phone, the attacker's job may be much more difficult to accomplish and require much longer, possibly resulting in the attack being discovered. Disabling the hub on the IP Phone will prevent this kind of attack. However, it is a rather simple task to turn the hub back on.

A.2 Integrity Issues

Integrity of information means that information remains unaltered by unauthorized users. For example, most users want to ensure that bank account numbers cannot be changed by anyone else, or that passwords are changed only by the user or an authorized security administrator. Telecommunication switches must protect the integrity of their system data and configuration. Because of the richness of feature sets available on switches, an attacker who can compromise the system configuration can accomplish nearly any other goal. For example, an ordinary extension could be re-assigned into a pool of phones that supervisors can listen in on or record conversations for quality control purposes. Damaging or deleting information about the IP network used by a VOIP switch results in an immediate denial of service.

The security system itself provides the capabilities for system abuse and misuse. That is, compromise of the security system not only allows system abuse but also allows the elimination of all traceability and the insertion of trapdoors for intruders to use on their next visit. For this reason, the security system must be carefully protected.

Integrity threats include any in which system functions or data may be corrupted, either accidentally or as a result of malicious actions. Misuse may involve legitimate users (i.e. insiders performing unauthorized operations) or intruders.

A legitimate user may perform an incorrect, or unauthorized, operations function (e.g., by mistake or out of malice) and may cause deleterious modification, destruction, deletion, or disclosure of switch software and data. This threat may be caused by several factors including the possibility that the level of access permission granted to the user is higher than what the user needs to remain functional.

Intrusion - An intruder may masquerade as a legitimate user and access an operations port of the switch. There are a number of serious intrusion threats. For example, the intruder may use the permission level of the legitimate user and perform damaging operations functions such as:

- disclosing confidential data
- causing service deterioration by modifying the switch software
- crashing the switch
- removing all traces of the intrusion (e.g., modifying the security log) so that it may not be readily detected

Insecure state - At certain times the switch may be vulnerable due to the fact that it is not in a secure state. For example:

- After a system restart, the old security features may have been reset to insecure settings, and new features may not yet be activated. (For example, all old passwords may have reverted to the default system-password, even though new passwords are not yet assigned.) The same may happen at the time of a disaster recovery.

- At the time of installation the switch may be vulnerable until the default security features have been replaced.

DHCP Server Insertion Attack

It is often possible to change the configuration of a target phone by exploiting the DHCP response race when the IP phone boots. As soon as the IP phone requests a DHCP response, a rogue DHCP server can initiate a response with data fields containing false information.

This attack allows for possible man in the middle attacks on the IP-media gateway, and IP Phones. Many methods exist with the potential to reboot the phone remotely, e.g. "social engineering", ping flood, MAC spoofing (probably SNMP hooks, etc.).

If possible, use static IP addresses for the IP Phones. This will remove the necessity of using a DHCP server. Further, using a state based intrusion detection system can filter out DHCP server packets from IP Phone ports, allowing this traffic only from the legitimate server.

TFTP Server Insertion Attack

It is possible to change the configuration of a target phone by exploiting the TFTP response race when the IP phone is resetting. A rogue TFTP server can supply spurious information before the legitimate server is able to respond to a request. This attack allows an attacker to change the configuration of an IP Phone. Using a state based intrusion detection system can filter out DHCP server packets from IP Phone ports, allowing such traffic only from the legitimate server. Organizations looking to deploy VOIP systems should look for IP Phone instruments that can download signed binary files.

A.3 Availability and Denial of Service

Availability refers to the notion that information and services be available for use when needed. Availability is the most obvious risk for a switch. Attacks exploiting vulnerabilities in the switch software or protocols may lead to deterioration or even denial of service or functionality of the switch. For example: if unauthorized access can be established to any branch of the communication channel (such as a CCS link or a TCP/IP link), it may be possible to flood the link with bogus messages causing severe deterioration (possibly denial) of service. A voice over IP system may have additional vulnerabilities with Internet connections. Because intrusion detection systems fail to intercept a significant percentage of Internet based attacks, attackers may be able to bring down VOIP systems by exploiting weaknesses in Internet protocols and services.

Any network may be vulnerable to denial of service attacks, simply by overloading the capacity of the system. With VOIP the problem may be especially severe, because of its sensitivity to packet loss or delay.

CPU Resource Consumption Attack without any account information.

An attacker with remote terminal access to the server may be able to force a system restart (shutdown all/restart all) by providing the maximum number of characters for the login and password buffers multiple times in succession. Additionally, IP Phones may reboot as a result of this attack.

In addition to producing a system outage, the restart may not restore uncommitted changes or, in some cases, may restore default passwords, which would introduce intrusion vulnerabilities. The deployment of a firewall disallowing connections from unnecessary or unknown network entities is the first step to overcoming this problem. However, there is still the opportunity for an attacker to spoof his MAC and IP address, circumventing the firewall protection.

Default Password Vulnerability

It is common for switches to have a default login/password set, e.g., admin/admin, or root /root. Similarly, VOIP telephones often have default keypad sequences that can be used to unlock and modify network information

This vulnerability would allow an attacker to control the topology of the network remotely, allowing for not only complete denial of service to the network, but also a port mirroring attack to the attacker's location, giving the ability to intercept any other conversations taking place over the same switch. Further, the switch may have a web server interface, providing an attacker with the ability to disrupt the network without advance knowledge of switch operations and commands. In most systems, telephones download their configuration data on startup using TFTP or similar protocols. The configuration specifies the IP addresses for Call Manager nodes, so an attacker could substitute another IP address pointing to a call manager that would allow eavesdropping or traffic analysis. Changing the default password is crucial. Moreover, the graphical user interface should be disabled to prevent the interception of plaintext administration sessions.

Exploitable software flaws

Like other types of software, VOIP systems have been found to have vulnerabilities due to buffer overflows and improper packet header handling. These flaws typically occur because the software is not validating critical information properly. For example, a short integer may be used as a table index without checking whether the parameter passed to the function exceeds 32,767, resulting in invalid memory accesses or crashing of the system.

Exploitable software flaws typically result in two types of vulnerabilities: denial of service or revelation of critical system parameters. Denial of service can often be implemented remotely, by passing packets with specially constructed headers that cause the software to fail. In some cases the system can be crashed, producing a memory dump in which an intruder can find IP addresses of critical system nodes, passwords, or other security-relevant information. In addition, buffer overflows that allow the introduction of malicious code have been found in VOIP software, as in other applications.

These problems require action from the software vendor, and distribution of patches to administrators. Intruders monitor announcements of vulnerabilities, knowing that many organizations require days or weeks to update their software. Regular checking for software updates and patches is essential to reducing these vulnerabilities.

Automated patch handling can assist in reducing the window of opportunity for intruders to exploit a known software vulnerability.

Account Lockout Vulnerability

An attacker will be able to provide several incorrect login attempts at the telnet prompt until the account becomes locked out. (This problem is common to most password-protected systems, because it prevents attackers from repeating login attempts until the correct password is found by trying all possible combinations.)

The account is unable to connect to the machine for the set lockout time. If remote access is not available, this problem can be solved with physical access control.

B   Appendix: VOIP Frequently Asked Questions

1. What is VOIP?

    Voice Over Internet Protocol is a set of software, hardware, and standards designed to make it possible to transmit voice over packet switched networks, either an internal Local Area Network, or across the Internet.

2. What are some of the advantages of VOIP?

    a. Cost – a VOIP system is usually cheaper to operate than an equivalent office telephone system with a Private Branch Exchange and conventional telephone service.

    b. Integration with other services – Innovative services are emerging that allow customers to combine web access with telephone features through a single PC or terminal. For example, a sales representative could discuss products with a customer using the company's web site. In addition, the VOIP system may be integrated with video across the Internet, providing a teleconferencing facility.

3. What are some of the disadvantages of VOIP?

    a. Startup cost – although VOIP can be expected to save money in the long run, the initial installation can be complex and expensive. In addition, a single standard has not yet emerged for many aspects of VOIP, so an organization must plan to support more than one standard, or expect to make relatively frequent changes as the VOIP field develops.

    b. Security – the flexibility of VOIP comes at a price: added complexity in securing voice and data. Because VOIP systems are connected to the data network, and share many of the same hardware and software components, there are more ways for intruders to attack a VOIP system than a conventional voice telephone system or PBX.

4. Can small organizations or home users use VOIP systems?

    Yes. Vendors have made VOIP solutions attractive to organizations of all sizes, and are now expanding into the home market. Most of the security problems discussed in this publication have also extended into the home market.

5. Can I use my existing network equipment (routers, hubs, etc.) for a VOIP network?

   No, except possibly for home use. VOIP demands high performance because voice communications must be in real time. A few seconds delay in data transmission is accepted and common, but a similar delay in a telephone conversation would make the system unacceptable to users. VOIP network equipment uses special protocols to over come the performance problems involved in transmitting voice over the Internet. Existing local area network cabling can be used.

6. Does VOIP require additional phone lines or new phone numbers.
   No, VOIP systems can be tied in to existing connections,

7. Can I use VOIP like a normal telephone?

   Yes, if the right hardware and software is installed. Some basic VOIP services are available from messaging programs such as AIM and Microsoft Netmeeting, but these programs cannot be used to dial telephone numbers. VOIP systems, coupled with a VOIP service provider, enable users to enter ordinary telephone numbers to make calls across the Internet.

8. Do both parties to a call need to have VOIP?

   No, VOIP service providers translate VOIP calls to and from a form that is carried on the conventional telephone network using IP/TDM gateways, so either party may use VOIP or conventional equipment.

9. How does the sound quality of VOIP compare with traditional systems?

   More recent VOIP systems have sound quality equivalent to conventional phones, especially if standalone VOIP phones are used. A "softphone" may have packet delay or other quality of service issues if used on a heavily loaded PC.

10. What is a VOIP "softphone"?

    The term softphone refers to a telephone capability implemented on an ordinary PC, using only special software and a microphone/headset that plugs into the PC's audio ports. As noted in the body of this publication, though, softphones should not be used where security or privacy are a concern because of the ease with which they can be attacked. These systems are also more vulnerable to denial of service attacks from worms and viruses.

11. What size PC is required to operate a softphone?

   A high-end PC is not needed. Most systems require only 64MB of RAM and a 200 MHz or faster processor. VOIP software is available for all popular operating systems.

12. Will a VOIP system continue to function during a power failure or cable outage?

   If all components have an uninterruptible power supply, the system should continue to function as long as the UPS batteries last. However, if the VOIP system is implemented on a cable modem, phone service will not be available during a cable outage. Similarly, if DSL is used, an outage of the DSL line will interrupt phone service. A conventional phone connection or mobile phones can serve as a backup.

C   Appendix: VOIP Terms

Application Level Gateway (ALG) – Application Level Gateways (ALGs) are application specific translation agents that allow an application (like VOIP) on a host in one address realm to connect to its counterpart running on a host in different realm transparently. An ALG may interact with NAT to set up state, use NAT state information, modify application specific payload and perform whatever else is necessary to get the application running across disparate address realms.

Abstract syntax notation one (ASN.1): A standard, flexible method that (a) describes data structures for representing, encoding, transmitting, and decoding data, (b) provides a set of formal rules for describing the structure of objects independent of machine-specific encoding techniques, (c) is a formal network-management Transmission Control Protocol/Internet Protocol (TCP/IP) language that uses human-readable notation and a compact, encoded representation of the same information used in communications protocols, and (d) is a precise, formal notation that removes ambiguities.

Call Processor – component that sets up and monitors the state of calls, and provides phone number translation, user authorization, and coordination with media gateways.

Codec – coder/decoder, which converts analog voice into digital data and back again, and may also compress and decompress the data for more efficient transmission.

Firewall Control Proxy - component that controls a firewall's handling of a call. The firewall control proxy can instruct the firewall to open specific ports that are needed by a call, and direct the firewall to close these ports at call termination.

H.323 - The International Telecommunications Union (ITU) standard for packet-switched network voice and video calling and signaling.

Jitter - non-uniform delays that can cause packets to arrive and be processed out of sequence

Latency – time delay in processing voice packets.

Media gateway - the interface between circuit switched networks and IP network. Media gateways handle analog/digital conversion, call origination and reception, and quality improvement functions such as compression or echo cancellation.

Media Gateway Control Protocol – common protocol used with media gateways to provide network management and control functions.

PSTN – the public switched telephone network.

QoS - Quality of Service - a network property that specifies a guaranteed throughput level.

Session Initiation Protocol (SIP) – A standard for voice and video over a packet switched network developed by IETF.

Sniffer – a network monitoring tool, usually a software tool running on a PC.

INDEX

advanced encryption standard (AES), 72
application level gateway (ALG), 59, 77
authentication, 58, 63, 68, 74, 76, 82
authentication header (AH), 63, 68, 73
Availability, 85
call manager, 3, 14, 53, 86
confidential, 81
Confidentiality, 81
cost, iv, 3, 16, 60, 76, 88
datagram, 64
DMZ, 4, 52
encapsulating security payload (ESP), 63, 65, 68, 73
encryption, 3, 17, 18, 19, 20, 24, 63, 64, 65, 66, 67, 68, 69, 72, 73
Federal Information Security Management Act (FISMA), 10
file transfer protocol (FTP), 52
firewall, 4, 20, 24, 36, 37, 52, 53, 56, 57, 58, 59, 60, 65, 73, 75, 83, 85
firewall control proxy (FCP), 59, 60
gateway, 26, 37, 52, 58, 64, 76
H.245, 26, 27, 28, 37
H.255, 26
H.323, 3, 5, 26, 27, 28, 36, 37, 39, 40, 57, 58, 59, 75, 76, 77, 79
HTTP, 52, 75, 82
HTTPS, 83
IKE, 63, 68, 73
Integrity, 83
Internet Engineering Task Force (IETF), 39, 60, 73
IP, 3, 4, 13, 15, 21, 23, 37, 45, 46, 52, 53, 54, 55, 56, 58, 59, 63, 65, 68, 69, 72, 73, 75, 78, 79, 80, 81, 82, 83, 84, 85, 86
IP phone, 53, 69, 83, 84, 85
IPsec, 4, 21, 24, 55, 62, 63, 64, 65, 66, 68, 69, 72, 73, 75, 78, 80
IPsec tunneling, 62, 65
IPv4, 73
IPv6, 56, 63, 65, 73, 80
ISP, 18, 54, 56, 80
ITU, 19, 26
jitter, 3, 19, 20, 21, 23, 57, 59, 65, 68, 72
latency, 3, 19, 20, 21, 22, 23, 24, 37, 57, 58, 59, 63, 65, 66, 67, 68, 72
local area network (LAN), 4, 23, 52, 54, 65, 69, 73
MAC, 4, 81, 83, 85, 86
middlebox, 37
NetMeeting, 13
network address translation (NAT), 4, 24, 28, 37, 40, 45, 46, 52, 54, 55, 56, 57, 58, 59, 60, 62, 68, 73, 75, 79, 80
Office of Management and Budget (OMB) Circular A-130, 10
packet loss, 3, 19, 21, 22, 23, 65, 67, 68, 73, 85
packet sniffer, 63, 65, 82
proxy, 37, 39, 45, 46, 53
proxy server, 37, 39, 45, 46, 53
public switched telephone network (PSTN), 13, 16, 18, 19, 22, 26
quality of service (QoS), 3, 19, 20, 21, 22, 23, 24, 28, 56, 57, 62, 63, 65, 66, 67, 68, 72, 73, 76, 78, 89
Real Time Protocol (RTP), 16, 20, 21, 23, 28, 40, 44, 45, 53, 56, 57, 58, 59, 76
registrar, 39, 46
router, 54, 56, 58, 64, 67, 69, 83
session initiation protocol (SIP), 3, 5, 26, 39, 40, 44, 45, 46, 57, 58, 59, 75, 76, 77, 79
stateful firewall, 37, 53, 58
TCP, 21, 22, 27, 37, 39, 45, 54, 65, 73, 85
UDP, 16, 20, 21, 36, 39, 45, 53, 54, 57, 68, 73
virtual private network(VPN), 57, 62, 63, 65, 75, 80
vulnerabilities, 10, 56, 81, 85, 86
wide area network (WAN), 4, 23, 52, 65
Windows, 13, 90

www.ingramcontent.com/pod-product-compliance
Lightning Source LLC
Chambersburg PA
CBHW081829170526
45167CB00007B/2758